# EASY AI

## MASTER THE BASICS OF ARTIFICIAL INTELLIGENCE, BUILD ESSENTIAL SKILLS, AND LEVERAGE AI FOR FUTURE SUCCESS!

### MICHAEL GRANT MALLOY

# CONTENTS

# INTRODUCTION

Imagine waking up on a Monday morning, coffee in hand, scrolling through a tidal wave of news about artificial intelligence break-throughs. Everywhere you look, online articles, podcasts, even water-cooler chats at work, the conversation is buzzing with AI jargon: machine learning, neural networks, generative models. Maybe you've felt that sinking sense of being left behind, wondering when or if you'll ever catch up.

You're not alone. Many savvy, hardworking professionals feel over-whelmed by the speed and complexity of AI's rise. But what if you didn't need a computer science degree to make these innovations work for you? What if, instead of feeling intimidated, you felt genuinely curious and confident about the possibilities ahead?

This book exists to answer those questions in the clearest, most prac-tical way possible. It's designed for people who want tangible career growth, higher productivity, and new income opportunities, but don't have time (or patience) for technical jargon or endless theory. Inside, you'll find hands-on strategies for using AI tools at work or in a side hustle immediately, with real-world examples you can relate to. My goal is for you to finish each chapter feeling more empowered and less anxious than when you started, ready to turn AI from a source of stress into a secret weapon for your professional life.

Skeptical? That's perfectly reasonable. There's a lot of hype out there, plenty of misinformation, and confusing chatter about what AI really is. Many books and articles assume you already know how to code or toss around buzzwords without explaining them. If you've ever tuned out because it all felt too complicated or contradictory, you're in good company. I've spoken with countless smart, motivated adults who admit they've avoided AI conversations out of frustration or self-doubt. But it doesn't have to be that way. This book starts with the belief that you are fully capable of mastering AI's core ideas, no math, no coding, no prior experience required.

Let's be clear: AI is no longer science fiction. It's a tool just like the internet, or smartphones. A tool that has become essential in our daily lives. AI is already simplifying things around you, from personalized playlists to smart home assistants. When used intentionally, AI can become a powerful assistant for managing tasks, solving problems, and sparking creativity.

At the same time, the world is flooded with AI myths and exaggerated promises. From headlines warning about robots stealing jobs to breath-less claims of overnight riches through AI, separating fact from fiction can be exhausting. That's why this book won't just explain how AI works, it will also explore when and why it matters in real workplaces. By offering honest answers to the most common questions and acknowledging where AI still falls short, I hope to earn your trust from page one. If you're tired of tech experts talking over your head, this book was written with you in mind.

Who am I, and why should you trust me as your guide? I've spent over a decade working in IT helping professionals across industries, from small business owners to Fortune 500 executives, make sense of emerging technologies. My background in data science and tech roll-out (education) is rooted in making complex ideas understandable for non-technical audiences. I've seen how a little clarity and the right tools can turn confusion into confidence, and anxiety into ambition. My passion is helping others discover how data science and AI can unlock new possibilities, both professionally and personally.

This journey isn't just about information, it's about transformation. It's about helping you build confidence, ask better questions, and use AI with purpose. We'll begin with the basics, clarifying what AI really is and busting common myths. Then, you'll learn how to identify and deploy tools that increase your productivity, help organize your day, or streamline projects without writing a single line of code. From there, we'll explore creative ways to monetize your skills using AI, while pausing along the way to examine the ethical dimensions of your choices. Each chapter builds on the last, so you never feel lost or rushed. You'll make steady progress and see real results with every step.

Education shouldn't feel like an obstacle course. That's why this book skips unnecessary technical deep dives and focuses instead on simplicity, clarity, and relevance. Complex terms are explained in plain English. Several chapters include practical tools and exercises, not homework, but real-life actions you can take to solidify your understanding and build confidence. Whether you prefer hands-on experimentation or thoughtful reflection, you'll find tools that suit your learning style.

I want to start by highlighting something that I initially thought was obvious, but my beta reader (my mom) pointed out that it might not be clear to everyone. You do not need to physically "go anywhere" to purchase these AI tools. Most of them can be found online, either on a computer or a mobile device. You can discover various AI tools through a keyword search on search engines like Google or Bing. I've also included lists of some of the most commonly used AI tools. In most cases, you will need to create an account on the platform or website where you find the AI tools you wish to use. If you're a non-commercial user, many of these tools offer free versions that you can access after setting up your account. For those AI platforms that do not have a free version, many provide a free trial period, allowing you to try out the tools before committing to a purchase.

Now that we got the not-so-obvious logistics out of the way, I'd like to point out that using AI isn't just about efficiency or profit. It's also about responsibility and critical thinking. With great technology comes great potential for misuse, from biased algorithms to privacy breaches. That's why we'll cover the deeper questions too: How do you spot bias in AI tools? How do you protect your data? How do you evaluate whether an AI system is being used ethically? These sections will help you develop healthy skepticism and ethical awareness, key ingredients for making AI work for you and those around you.

The future doesn't wait for anyone. But with the right approach, you can ride the next wave of technological change instead of struggling to stay afloat. Consider this book not just a guide, but a launchpad for new habits, skills, and perspectives. Each page is designed to build your confidence, ignite curiosity, and reframe AI as a tool of empowerment not intimidation. Whether you're focused on career growth, launching a business, or simply staying informed, you're about to discover how practical and powerful AI can truly be.

So, let's get started. Bring your questions and skepticism, your hopes and doubts. Together, we'll cut through the noise and uncover a clear, confident path to practical, profitable, and ethical AI usage designed for real people living in the real world. By the end of this journey, you'll have the tools and mindset to seize the opportunities AI offers today and, in the years, to come. Welcome to your AI learning adventure.

# SECTION ONE: CONCEPT

# CHAPTER 1: GETTING STARTED WITH ARTIFICIAL INTELLIGENCE

WHEN PEOPLE THINK of artificial intelligence, they often picture a futuristic robot with endless knowledge and human-like thoughts, something straight out of a science fiction movie. Reality, while less dramatic, is far more compelling. AI is already deeply embedded in our everyday lives, often in ways we don't even notice. It powers the virtual assistant that sets your alarm, the recommendation system that suggests your next favorite song, and the navigation app that helps you avoid traffic. While these tools may seem almost magical, they are not driven by independent thought they operate through data analysis, pattern recognition, and carefully designed programming.

It's easy to overlook how much AI influences daily activities, both at home and in the workplace. Whether it's an email filter organizing your inbox, a doctor using AI to detect a rare condition early, or a package arriving sooner than expected thanks to optimized logistics these are just a few examples of AI at work. As more industries and roles adopt AI tools, understanding what AI is, how it functions, and where its limitations lie becomes essential not just for engineers or tech experts, but for everyone. Gaining this knowledge opens the door to smarter problem-solving, time-saving innovations, and even new career opportunities.

## DEMYSTIFYING ARTIFICIAL INTELLIGENCE: CORE CONCEPTS, EVOLUTION, AND EVERYDAY IMPACT

Artificial intelligence (AI) refers to computer systems designed to perform tasks that have typically required human intelligence. These systems can analyze large amounts of data, recognize patterns, and make decisions using algorithms, which are step-by-step instructions programmed by humans. For instance, voice recognition technology works by analyzing speech patterns and matching them with known words and commands. When you ask a virtual assistant a question, it uses AI to interpret your speech, understand your intent, and provide a response. Another common application is image processing, where AI can detect objects in photos, recognize faces, and categorize scenes based on visual information.

AI systems achieve these abilities through a combination of learning from existing data and following specific rules set by programmers. One key approach is machine learning, a subset of AI in which computers learn from examples rather than relying solely on fixed instructions. This means an image-processing AI can improve over time as it encounters more pictures and learns to recognize new patterns. A more advanced form, deep learning, uses artificial "neural networks" inspired by the human brain's ability to solve problems. These computer networks imitate the brain by pulling information from millions of resources on the internet (brain cells) to provide the appropriate answers to whatever was asked of it. These neural networks allow AI to perform highly complex tasks, such as language translation or diagnosing diseases from medical images.

The idea of AI predates modern smartphones and smart speakers. In the 1940's, British mathematician Alan Turing proposed that machines could simulate human reasoning, laying the foundation for the field of artificial Intelligence. His famous Turing Test aimed to determine whether a computer could respond in conversation so convincingly that a human couldn't tell the difference. This idea stretched the boundaries of what machines might one day accomplish.

In the 1950s and 1960s, researchers explored symbolic AI, where computers used symbols and logic to solve problems like chess or mathematical puzzles. The term *artificial intelligence* was coined during this period at a 1956 research conference at Dartmouth College, marking the formal birth of the field. In the 1970s and 1980s, expert systems emerged programs designed to replicate the decision-making abilities of human specialists in fields such as medicine and finance. While early AI had significant limitations, it laid the groundwork for future breakthroughs.

From the 2000s onward, AI research made rapid strides, especially in deep learning and neural networks. These techniques enabled computers to process vast amounts of data with greater accuracy than earlier models, driving advancements in real-time language translation, facial recognition, and image analysis.

Popular culture has often portrayed AI as conscious, self-aware robots or super intelligent machines bent on world domination. In reality, current AI systems lack emotions, consciousness, and independent will. They cannot think or feel like humans. Instead, today's AI systems operate strictly within the boundaries set by their programming and data.

Even the most sophisticated AI today doesn't possess self-awareness or the ability to make decisions on its own. It acts according to its design, lacks motives, and does not make decisions outside its coded parameters.

Understanding this distinction is crucial. While AI is powerful and transformative, it is not a conscious being. This means it is not aware and does not think on its own. It requires input from a human operator to perform any type of task. Knowing AI's strengths and limitations allows us to use it more effectively and responsibly in the world around us.

In daily life, people interact with artificial intelligence more often than they realize. Virtual assistants like Siri, Google Assistant, and Alexa use natural language processing (a branch of AI) to interpret speech and provide helpful responses. Whether you're checking the weather, playing music, or setting a reminder, these assistants analyze your commands and respond accordingly.

Streaming platforms such as Netflix and Spotify also rely on AI-driven recommendation systems to suggest shows or songs based on your past behavior. The more you engage, the better the suggestions become, thanks to machine learning algorithms that adapt to your preferences over time.

Customer service has also been transformed by AI. Many companies now deploy chatbots and automated messaging systems to address inquiries, troubleshoot issues, and guide users through purchases. These systems handle routine tasks efficiently, allowing human agents to focus on complex problems. In online shopping, AI personalizes experiences by recommending items similar to those you've browsed or bought, aligning with your interests.

In healthcare, AI helps doctors interpret medical images, assess patient risk, and develop personalized treatment plans. In transportation, AI powers navigation apps that analyze real-time traffic data to suggest the quickest routes and predict delays. These practical applications enhance convenience, safety, and efficiency in everyday routines.

This foundational understanding of AI sets the stage for deeper exploration of its core technologies and ethical considerations later in the book. In Chapter 2, we'll dive into key concepts like machine learning and neural networks, while Chapter 4 will examine the ethical dimensions of AI. For now, let's continue exploring its practical value and potential.

**Key Types of AI, Simply Explained**

There are several types of AI, but here are the ones you'll encounter most often:

• **Narrow AI** – The most common type. It does *one thing well* (like facial recognition or language translation).
*Example: ChatGPT answering your questions, or Google Maps finding the fastest route.*

- **Machine Learning (ML)** – A type of AI that improves through experience. It learns from data without being explicitly programmed.
  *Example: Netflix recommending shows based on what you've watched.*

- **Neural Networks** – A kind of machine learning inspired by the human brain's structure. Great for complex tasks like image or speech recognition.
  *Example: Facebook auto-tagging faces in photos.*

- **Generative AI** – A newer form of AI that can create new content—like text, images, or even music—based on prompts.
  *Example: DALL·E generating art from your description, or ChatGPT writing an email draft.*

## THE REAL VALUE OF AI: DATA, PRODUCTIVITY, AND CAREER GROWTH, AND REAL-WORLD SOLUTIONS

From home studios to corporate offices, AI is becoming an indispensable tool in daily workflows. Professionals across industries use AI to manage customer chats, automate billing, schedule meetings, draft emails, and generate social media content. These tools streamline repetitive work, freeing up time for creativity and strategic thinking. For example, a freelance marketer might use AI to scan analytics and recommend the best times to post ads saving hours each week and enabling more client interaction. As AI becomes woven into everyday work, it opens opportunities to earn extra income by offering faster turnaround or new services using these technologies.

AI's strength lies in processing vast amounts of data quickly, at speeds far beyond human capability. In retail, AI swiftly sifts through sales records, online reviews, and web traffic to spot trends, such as which products will likely be popular during next season's holidays. This enables store managers to adjust inventory proactively. Medical professionals use AI systems that review thousands of patient files in minutes to recognize signs of rare diseases. By detecting subtle patterns humans might overlook, AI helps doctors make earlier and more accurate diagnoses, improving survival rates. In both cases, companies and workers make better decisions because AI turns raw data into actionable insights at lightning speed.

Organizations of all sizes benefit from these insights. AI that prioritizes incoming emails or verifies loan applications can save administrative hours. Banks now approve some loans within hours instead of days, thanks to automated processing. In warehouses, AI-driven robots and logistics software optimize delivery routes, cut costs, and boost efficiency. These advancements improve productivity and job satisfaction while conserving valuable resources.

New career paths are opening for individuals without deep technical backgrounds. Many companies hire annotation specialists whose job is to label images and flag errors, helping AI systems "learn" correct outputs. Content moderators use AI monitoring tools to filter inappropriate posts or comments online, which supports safe and positive digital spaces. Another emerging role is prompt engineering, where workers design effective questions or instructions for AI chatbots and virtual assistants. These positions don't require advanced programming skills but rely on clear communication, attention to detail, and problem-solving. Freelancers also profit by using AI in their side hustles, like running automated resume-writing services or curating personalized reading lists for customers. With certifications or short courses, for example, micro-credentials in data visualization, workers can quickly build skills and qualify for these new roles.

AI isn't just about saving time, it's solving challenges that impact health, safety, and the environment. In healthcare, AI detects early stages of diseases like cancer by analyzing medical scans faster and more consistently than traditional methods. At one hospital, integrating AI-supported screening led to a 10 percent increase in early cancer detection rates. Shipping and logistics companies use AI to analyze traffic, weather, and package data, choosing optimal delivery routes that reduce miles driven and fuel used. One major carrier cut average delivery times by 20 percent after deploying AI route-planning software. Environmental groups employ AI-powered cameras and sensors to track animal movements in protected areas, which helps stop poaching and maintain healthy wildlife populations. Each success story illustrates how AI addresses real-world problems, protecting lives and natural resources while optimizing human effort.

As we'll see in Chapter 5, these productivity gains and new income opportunities become tangible through specific AI tools and strategies. Chapter 6 will show how non-technical users can easily leverage AI tools, while Chapter 10 will teach prompt mastery to enhance AI effectiveness.

The practical side of AI is all around us, shaping decisions in work and daily life. Whether it's helping doctors diagnose illnesses sooner, making online shopping more personal, or creating new career options for people from different backgrounds, AI is becoming an everyday tool. As you continue through this book, you'll keep seeing how these technologies help solve real problems and open up opportunities. With a clear foundation now, you're ready to discover more ways AI can support both your goals and the world around you.

### What Is AI—Really?

Artificial intelligence, or AI, refers to computer processes designed to perform tasks that have typically required human intelligence. These tasks include things like:

- Understanding language (reading, writing, translating)
- Recognizing images or voices
- Solving problems
- Making decisions
- Learning from past experiences

At its core, AI is designed to simulate certain types of human thinking. But it doesn't do this by "thinking" the way we do—it analyzes patterns in vast amounts of data and predicts what response, or action makes the most sense.

### What AI Is *Not*

- It's not conscious or self-aware.

- It doesn't have feelings or personal opinions.
- It doesn't understand things the way humans do—it predicts based on patterns in data.

Think of AI as an incredibly powerful assistant. It can be creative, fast, and surprisingly helpful—but it still needs human guidance.

# CHAPTER 2: BREAKING DOWN AI: THE CORE PILLARS EXPLAINED

WHETHER IT'S a streaming service predicting your next favorite show or your phone unlocking with a glance, artificial intelligence (AI) seems to be everywhere. Yet, how these systems actually work often feels mysterious, almost magical. It's easy to use these tools without a second thought about how they understand us, sort through massive amounts of data, or recognize voices and images. Most people interact with AI daily yet rarely pause to explore what's happening behind the scenes.

This book aims to pull back that curtain and provide clear, relatable explanations of the core ideas that make AI function. Through simple stories and real-life comparisons, we'll uncover the everyday logic powering intelligent machines. From recognizing patterns to making decisions in human-like ways, you'll gain a better understanding of how artificial intelligence shapes the world around you.

# MAKING MACHINE LEARNING UNDERSTANDABLE

There are three main types of machine learning: supervised, unsupervised, and reinforcement learning. Each works differently but connects to everyday life in ways many people don't realize.

Supervised learning is like teaching with flashcards. Imagine showing a child pictures of animals while naming each one: "This is a cat. This is a dog." Eventually, the child learns to identify cats and dogs without help. For computers, this involves training on large sets of labeled data, for example, thousands of images tagged as 'cat' or 'dog.' With enough examples, the machine can correctly sort new images based on what it has learned.

Unsupervised learning, on the other hand, works without labeled data. It's more like sorting a mixed basket of fruit by noticing similarities grouping apples because they're round and red, bananas because they're long and yellow. Businesses use this technique to uncover patterns in customer behavior, such as identifying groups of shoppers who frequently buy organic items. The system finds these clusters on its own, without being told what to look for.

Reinforcement learning is based on trial and error, much like training a puppy. When the puppy sits and gets a treat, it learns to repeat the behavior. If it jumps on the couch and gets ignored, it eventually stops. Computers can learn the same way by receiving rewards or penalties based on their actions. For example, a program might play thousands of games of chess against itself, learning better strategies each time it wins, until it becomes a master.

Many people encounter machine learning every day, often without realizing it. One common example is the email spam filter. At first, it learns from messages marked as *'spam'* or *'not spam.'* Over time, it identifies key features like suspicious subject lines or unfamiliar senders and gets better at catching unwanted emails before they reach your inbox. The more you interact with it, the smarter it becomes, adapting to new tricks used by scammers.

Streaming services also rely heavily on machine learning. Each time you watch, skip, or rate a show, the system records your preferences. By comparing your habits with millions of others, it spots patterns for instance, you might favor romantic comedies set in Italy or finish every documentary about space. These insights help generate personalized recommendations that feel surprisingly accurate.

Virtual assistants, like those on smartphones or smart speakers, learn in similar ways. When you ask for the weather, set alarms, or look up facts, the assistant remembers your preferences and speaking style. Over time, it improves its voice recognition and adapts to your routine, offering faster and more relevant help.

In the workplace, machine learning helps automate repetitive tasks, such as extracting details from invoices or matching receipts. This reduces human error and frees up time for more creative or strategic work. In healthcare, doctors use machine learning tools to detect diseases earlier by comparing new test results with vast databases of past cases. These systems highlight unusual patterns, aiding in quicker and more accurate diagnoses.

Banks use machine learning to detect fraud. Each transaction is analyzed to see if it fits your usual spending habits. If something appears suspicious, like a purchase made in another country, or an unusual amount, the system flags it. As it learns more over time, it becomes faster at spotting threats while reducing false alarms.

Building on the foundational ideas explored in Chapter 1, where we introduced the wide-ranging applications of AI, machine learning stands out as a central technology driving much of that progress. Understanding the three main types supervised, unsupervised, and reinforcement learning gives us a clear view of how machines learn from experience. With this knowledge in hand, let's now explore another core concept in AI: neural networks and deep learning.

## DEMYSTIFYING NEURAL NETWORKS

Neural networks are the next core pillar of AI. Neural networks build on the concept of machine learning by structuring learning in ways that mimic the human brain. Imagine your brain at work: when you see a familiar face or recognize a favorite song, countless brain cells, called neurons, work together, passing signals between each other. Neural networks operate similarly. They use artificial "neurons," which are called nodes in Tech language. Each node receives information, processes it, and passes its findings to the next set of nodes. These nodes are connected by pathways, similar to synapses in the brain. When a particular pathway proves especially effective for arriving at the correct answer, the connection becomes stronger.

Another way to think of these layers is like teams in a company. The first team handles raw input. If this were a group of people sorting mail, that first team would separate letters from packages. The next team layer builds upon those insights, identifying more complex patterns, like zip codes or destinations. Each subsequent layer combines what came before, finding more meaning at each stage, maybe adding overall milage or distance to destination. Just like how our brains process simple sensations first and then integrate them into a bigger understanding, neural networks follow this same principle: they progress from basic to complex information, one layer at a time.

## DEEP LEARNING

Deep learning expands on the concept of neural networks by stacking on even more layers of nodes, hence the term deep. This architecture allows machines not only to detect surface-level features but also to uncover intricate, hidden relationships in data. It's like learning to read between the lines of a story: deep learning captures subtle patterns that simpler models often miss. Take facial recognition, for instance, when you unlock your phone with your face, it's not just detecting your eyes or mouth but analyzing dozens of tiny details across several layers to ensure a precise match.

Let's bring this to life with real-world examples. Every time you're tagged in a social media photo, deep learning powers the image recognition feature. The system looks at pixels and patterns, compares them across millions of examples, and matches faces with names. Another practical case is customer service chatbots. When you ask a question about your bank account, a chatbot uses deep learning to interpret your words and figure out the best response, whether you want to check your balance or report an issue. Even voice assistants, like those sitting in smart speakers or on your smartphone, rely on these techniques. They listen to your request, break down your speech, and learn your preferences over time, improving accuracy with more usage. Self-driving cars are perhaps the most exciting example. They constantly analyze their surroundings, identifying traffic lights, pedestrians, and road signs in real time. Layers of neural networks help them safely steer through complicated scenarios, adapting quickly to new obstacles.

One of deep learning's biggest strengths is its ability to manage enormous volumes of information. Unlike traditional systems that required human experts to define features manually, deep learning learns directly from the data. This capability makes it especially valuable in complex situations where there's too much detail for any person to program step by step, like picking out suspicious payment patterns in online transactions or translating spoken language instantly.

Importantly, you don't need a degree in computer science or advanced math to benefit from neural networks and deep learning. These technologies are already embedded in everyday tools and apps organizing photos by location, suggesting friends or songs, managing emails, and more. In the workplace, they automate routine tasks, reveal insights, and support smarter decisions. Even in personal tools, like health-tracking apps or budgeting platforms deep learning is working behind the scenes to streamline and improve daily life.

From powering smart assistants to driving innovation in healthcare, finance, and entertainment, deep learning now plays a crucial role in making our world more efficient and intelligent. We'll revisit its applications in personal growth and the workplace in Chapters 7 and

8. For now, let's break down how neural networks operate and why deep learning has become such a game-changer.

The difference between traditional software and the neural networks operating AI comes down to flexibility, learning, and behavior. Here's a simple breakdown:

| Feature | Traditional Software | Artificial Intelligence |
|---|---|---|
| Behavior | Follows fixed, pre-written rules | Learns from data and adapts over time |
| Programming style | "If this, then that" logic | Uses algorithms that learn from examples |
| Improves over time? | No – only updates with new programming | Yes – improves through experience and data |
| Handles uncertainty? | Poorly – needs clear instructions | Well – can interpret fuzzy or ambiguous input |
| Example | Calculator, alarm clock app | ChatGPT, Google Translate, Netflix suggestions |

Although the science behind AI systems is highly advanced, you don't need technical expertise to benefit from them. Today's devices, apps, and services integrate AI seamlessly, operating quietly in the background to enhance efficiency and unlock new possibilities for work and play. Developing a basic understanding of these technologies should help you feel more confident in navigating the growing role AI plays in everyday life.

## FURTHER RESOURCES ON NEURAL NETWORKS AND DEEP LEARNING

For those of you who are always wanting more, needing to dig deeper, tapping into expert-backed resources can help reinforce your insights and open new avenues for exploration into this amazing new world. Below you'll find a handpicked list of resources reflecting a variety of perspectives—academic research, industry breakthroughs, educational guides, and real-world analyses. These resources are not meant for everybody, only those knowledge-seekers who can never get enough!

Together, the references below will deepen the foundation you've developed, affirm major points covered in this chapter, and provide clear paths for continued learning.

**Foundational Overviews and Academic Analyses**

1. **Goodfellow, I., Bengio, Y., & Courville, A. (2016). Deep Learning.** This widely acclaimed textbook offers a comprehensive, yet approachable, introduction to machine learning and neural networks. It's referenced by experts and newcomers alike for its clear explanations on how neural architectures learn complex patterns, supporting earlier chapter claims about the versatility and adaptability of neural systems.
2. **LeCun, Y., Bengio, Y., & Hinton, G. (2015). "Deep learning." Nature, 521(7553), 436-444.** Written by pioneers in the field, this article summarizes the key developments that shaped modern deep learning—helpful for readers seeking authoritative context around the evolution and impact of these techniques.
3. **MIT OpenCourseWare – Introduction to Deep Learning (2023 update).** Offering free video lectures and course notes, MIT provides engaging introductions to neural architectures, backpropagation, and practical training strategies, making it suitable for non-technical readers aiming to bridge theory with application.

**Industry Whitepapers and Expert Blogs**

4. **Google AI Blog: "A Primer on Deep Learning" (2022).** Google's engineers break down neural networks, data handling, and real-world implementation tips. This resource links directly to hands-on projects and showcases the technology's reach across industries—a direct extension of the practical examples discussed previously.
5. **OpenAI Whitepaper: "GPT-3: Language Models are Few-Shot Learners" (2020).** Highlighting state-of-the-art advances

in large neural models, this whitepaper demonstrates how neural nets achieve impressive performance without vast retraining. It supports chapter points on scalability and generalization of deep learning systems.

6. **Microsoft Research: "Deep Residual Learning for Image Recognition" (He et al., 2016).** This paper introduces so-called 'ResNet' models, critical for solving complex tasks like modern image classification, tying back to the discussion on architecture innovations that make networks more powerful and robust.

## Educational Platforms and Accessible Guides

7. **DeepLearning.AI's Coursera Specialization (2023).** Developed by Professor Andrew Ng, this series covers neural network basics through advanced deep learning topics, making AI accessible for learners from any background. The modular lessons reinforce concepts featured earlier and include interactive exercises for practical understanding.

8. **Khan Academy: Neural Networks (2023).** Designed for beginners, Khan Academy's easy-to-follow tutorials focus on core mechanics—like neurons, weights, and activation functions—with visual aids that clarify challenging ideas introduced during your earlier reading.

## Specialized Research and Data-centric Perspectives

9. **Stanford CS231n: Convolutional Neural Networks for Visual Recognition (2023 lecture notes).** Used in one of the world's top AI courses, these notes explore how neural networks power advances in computer vision, adding depth to the architectural discussion while delivering up-to-date code samples and datasets for experimentation.

10. **DeepMind: "AlphaFold: Using AI for Scientific Discovery" (2021 report).** This resource analyzes AlphaFold's game-changing neural model for protein folding, demonstrating real-world scientific breakthroughs made possible by deep

architectures and echoing earlier points about cross-disciplinary applications.

11. **Fast.ai Practical Deep Learning for Coders Course (2023).** Fast.ai takes a bottom-up approach, focusing on hands-on coding and immediate results. Their curriculum enables faster iteration and tangible progress, ideal for those inspired to experiment after learning about foundational principles.

# CHAPTER 3: DISPELLING THE MYTHS: WHAT AI CAN AND CANNOT DO

THINK AI can do anything a person can do? That idea has traveled fast, but it only tells half the story. There's a lot of talk about smart machines taking over, changing how we work, and replacing people left and right. But much of what people believe about AI comes from hype, not reality. Some limits are built into how these systems work— and many supposed breakthroughs turn out to be less magical when you look closely.

If you've ever found yourself confused by AI marketing or worried about the future of your job, you're not alone. Sorting through all the claims can feel overwhelming. With so many buzzwords floating around, it's easy to miss what's really possible and where human skills remain vital. This chapter aims to shine a light on what AI actually does well, where it falls short, and how to make sense of all the noise.

## UNDERSTANDING AI'S LIMITATIONS AND CORRECTING MISCONCEPTIONS

AI systems have made remarkable strides, but there are clear boundaries to what they can do. One of the most important limitations lies in tasks involving genuine creativity and emotional intelligence. Although AI can generate poetry, music, or visual art, it does so by identifying and replicating patterns from vast datasets. It cannot bring fresh ideas out of nowhere or create from personal experience or emotional depth. For example, when AI composes music, it mixes elements learned from thousands of songs, so the result often lacks genuine emotion or novel style that characterizes human innovation. AI also struggles to understand the context and subtlety behind emotions. If you share a personal story with an AI chatbot, its responses will be based on word associations rather than any real empathy or intuition. This makes it easy for misunderstandings to occur. Complex decision-making further highlights this gap. Humans weigh unspoken cues, values, and potential consequences; AI simply sorts probabilities. Quality data and clearly defined goals are essential for the performance of AI systems. Inaccurate, limited, or biased datasets can lead to poor AI performance, sometimes resulting in serious consequences.

For instance, early facial recognition systems failed to recognize people of color at much higher rates because the training data lacked diversity. Similarly, if an AI system used in healthcare receives unclear instructions or relies on incomplete patient records, its recommendations may overlook dangerous conditions or propose ineffective treatments. During the COVID-19 pandemic, some diagnostic AIs saw their accuracy drop dramatically when applied to new data outside their training set, making them unreliable for real-world use cases. Unclear objectives or poorly communicated goals by the user(s) contribute to this problem. Without specific targets, AI can unintentionally optimize the wrong outcomes. For instance, recruitment tools have been known to unfairly disadvantage resumes associated with women due to historical biases present in past hiring data.

## THE TRUTH ABOUT AI AND JOBS

A common misconception is that AI will eventually replace most, if not all, human jobs. The reality is more complex. AI excels at automating routine, repetitive tasks like organizing emails or managing schedules but it works best as a partner, not a replacement. In healthcare, for instance, AI helps doctors analyze images and spot patterns, but final diagnoses still depend on medical expertise to weigh factors the AI cannot see. Far from wiping out employment, AI is helping to reshape the current and future job market. New roles are emerging, such as AI prompt engineers, data ethics officers, and algorithm auditors. These jobs require human strengths like critical thinking, ethics, and interpersonal skills. Rather than leading to massive unemployment, AI is encouraging businesses to upskill workers and opening opportunities in areas like ethical oversight, risk management, and creative strategy.

Some tasks remain firmly rooted in human capabilities. Problem-solving situations that require judgment under uncertainty, navigating ambiguous scenarios, or balancing competing moral dilemmas are areas that require essential human insight. In finance, for example, AI might catch unusual spending patterns, but only a person can interpret those signals in the broader context of economic uncertainty or customer intent. In art and storytelling, human creators infuse their work with meaning drawn from culture, history, and individual experience, qualities not replicable by statistical models alone. Handling sensitive conversations, negotiating, or mediating disputes also depends on reading emotions and shifting strategies according to subtle social cues, something AI still cannot manage reliably. Ethical decision-making, such as weighing privacy risks, fairness in hiring, or medical treatment priorities, requires values-based choices that demand ongoing human oversight.

## HUMAN-AI COLLABORATION: A POWERFUL PARTNERSHIP

When people use AI strategically, in other words, when people and AI work together, the results can be transformative. AI contributes speed and analytical power, while people bring creativity, empathy, and ethical reasoning. This partnership turbocharges productivity and enables breakthroughs that neither could accomplish alone.

As we explored in Chapter 2, foundational AI concepts like machine learning and neural networks provide the technical backbone for these capabilities, as well as any limitations. Understanding these core pillars helps clarify what AI can realistically achieve. While AI excels at pattern recognition and data processing, human skills remain essential for creative thinking, emotional intelligence, and ethical oversight. As we move forward, recognizing the real value of AI involves focusing on how people and technology together can unlock new possibilities, guided by a clear-eyed view of both strengths and weaknesses.

## EVALUATING AI PRODUCTS AND CLAIMS: A PRACTICAL APPROACH

Whether integrating AI into your business or just trying to understand how AI could improve your daily life, when you're paying for AI technology, learning to spot exaggerated marketing claims is essential. We'll expand on this in Chapter 6 when we explore user-friendly AI tools, but the fundamentals apply here too. Spotting overpromises in AI product claims requires a keen eye for red flags. When vendors describe their tools as 'intelligent,' 'fully autonomous,' or claim to provide 'state-of-the-art' capabilities without explaining the details, be cautious. For example, generic statements like 'powered by advanced deep learning' sound impressive but mean little unless backed by the actual outcomes. Excessive reliance on buzzwords, machine learning, neural networks, predictive analytics, without clear explanations often signals inflated promises. Real-world incidents have highlighted these risks: the 2010 Flash Crash, largely attributed to unchecked algorithmic trading, demonstrated how misplaced trust in opaque automation can lead to severe market disruptions. Sometimes, companies tout transformative fraud detection or recruitment tools that are actually built on basic rule-based coding, not real AI.

To identify missing information about the limitations of a specific AI product, look for a lack of discussion regarding human input required, data used to create or "train" the AI product, or the error rates. If an AI product seems to promise perfection without admitting any boundaries, that's a serious warning sign. Watch for vague claims about "learning" or "adapting," especially if no technical documentation supports how that happens.

Measuring value starts with aligning AI tools to your actual needs. Begin by defining clear objectives: What problem should the AI solve? For instance, an HR department might seek AI to screen resumes faster. In this scenario, measure potential solutions by how easily they fit existing HR processes. Does the tool plug into your current systems, or will it require extensive adjustment? Next, consider usability and scalability. Will metrics such as time to onboard, the learning curve for staff, number of transactions handled per hour, or increase in process accuracy, give tangible evidence of value. Case studies will help illuminate success and failure.

- **Successful example**: A retail company implemented AI-powered inventory software and reduced stockouts by 30% within six months.
- **Unsuccessful example**: A company abandoned an expensive chatbot because staff couldn't adjust its behavior, and it consistently misinterpreted customer inquiries.

When evaluating AI products to measure if they are having a meaningful impact, track outcomes such as personal cost or time savings, or for a business, staff productivity improvements, revenue growth, and / or user satisfaction.

## CUTTING THROUGH MARKETING HYPE

Analyzing marketing claims is essential to cut through noise and find real potential. Terms like 'next-generation AI,' 'self-learning system,' or 'automated intelligence' often appear in promotional materials. Break down these terms: does 'NEXT GENERATION' refer to an actual improvement, such as a new algorithm, or is it simply a colorful makeover of a outdated user dashboard that is still based solely on fixed rules?

To verify marketing claims:

- Check independent reviews from reputable tech analysts, websites, or publications.
- Compare promotional language with documented features and real capabilities. In other words, try before you buy! Most products will offer a free trial. If they don't, beware!

- For complex and/or expensive AI products, primarily used in business or complex industries, request the developers' published benchmarks, technical white papers, or case studies.

For example, a vendor might claim "seamless integration across platforms," but third-party reviews may highlight compatibility issues or high maintenance costs. Always look for specific use cases, performance metrics, and critical feedback, not just polished sales decks.

## ASKING THE RIGHT QUESTIONS

Evaluating an AI product requires evidence that it will work as promised. Ask vendors the following:

- What data does your AI use, and how is it sourced?
- How is the model validated?
- Are there independent audits or certifications?
- How is the AI retrained to prevent model-drift over time?
- What are the documented failures or limitations?
- Evaluate case studies provided by the vendor—not only for positive outcomes but also for transparency about challenges and failures. Reliable vendors include measured results, control group comparisons, or long-term follow-ups in their documentation.
- Inspect third-party reviews on neutral platforms or solicit opinions from peers at similar organizations who already use the product. This peer validation can uncover hidden support issues or false claims of ROI.

Whenever possible, insist on:

- Hands-on demos,
- Pilot projects, or
- Limited-time trials

Also, insist on tracking specific results during these phases. If the vendor can't measure outcomes, beware!

AI is only truly useful when it is reliable, transparent, and well-matched to the user's actual needs. Responsible selection ensures that businesses and individuals avoid costly mistakes, protect sensitive data, and incorporate genuinely efficient and useful AI tools into their life and work, rather than chasing empty promises or exposing themselves to risks. While AI is a powerful enabler, it's not a cure-all. Its value emerges when paired with human insight, creativity, and ethical reasoning areas where people continue to lead.

Whenever bold claims about new AI tools arise, move forward with curiosity and skepticism. Ask smart questions. Seek proof. Avoid being dazzled by buzzwords. Ultimately, combining practical evaluation with an honest appraisal of AI's capabilities will help avoid costly mistakes. The most effective strategy is recognizing where technology can assist you, and where your own skills matter most.

# CHAPTER 4: ETHICS AND RESPONSIBILITY: NAVIGATING AI'S DILEMMAS

A FEW YEARS AGO, I watched a new AI-powered app go viral almost overnight. People everywhere were using it for everything—writing emails, picking vacation spots, even making financial choices. My friends and colleagues raved about how easy life had suddenly become. It was exciting, but there was an undercurrent of worry as small news stories started popping up about users leaving sensitive information exposed or getting caught off-guard by odd mistakes made by the system. These early days felt like standing on the edge of something huge, both fascinating and uncertain, and conversations quickly shifted from *"look what it can do"* to *"should we be trusting this?"*

Not long after that, I realized many people, myself included, were facing a new set of questions. It wasn't just about making AI do what we wanted; it was about understanding our role in shaping its impact. This chapter is a chance to pause and think about the choices we make when working with these powerful tools. Together, we'll look at where things can go wrong, and most importantly, how we can use AI responsibly so that it helps and protects everyone involved.

# RESPONSIBLE AI USE GOES BEYOND PROMPT CRAFTING

The primary way people interact with AI tools is by giving instructions to the AI tool, either in verbal or written form. If you have an Amazon Alexa, think about how you elicit the information you want from this tool. If you ask it a relatively simple or vague question, you'll get a relatively simple or vague response. For example, your son has an evening soccer game today, so you ask Alexa *"What is the weather forecast is for today?"* You'll get a broad-based answer, such as *"Mostly sunny, with a 30% possibility of showers."* If you ask *"Alexa, what is the weather forecast is for Anytown, USA, between approximately 5:00 PM to 7:00 PM?"* Alexa may tell you that, *"while there is a 30% chance of showers today, the showers are expected to pass your area between noon and 2:00PM, and the rest of the day and evening is expected to be clear skies."*

What has become clear is that carefully tailored instructions (inputs), otherwise known as prompts, lead to more accurate and relevant responses (outputs) from AI systems. The process of creating instructions tailored to elicit specific information is often referred to as Prompt Engineering or Prompt Crafting. While mastering prompts improves AI outputs, ethical responsibility requires even deeper thought when crafting the instructions we give our AI tools, particularly in the business world. Technical skill alone is not enough. Responsible users must actively address privacy protection, bias recognition, transparency assessment, and personal accountability whenever they engage with AI tools.

## 1. Privacy Protection

Privacy is a cornerstone of ethical AI use. AI applications often collect a wide range of personal data, location, device details, usage habits, search history, and even inferred emotional states. Some tools also access names, payment information, contact lists, and behavioral preferences.

To safeguard this data:

- Review the tool's privacy policy thoroughly.
- Look for clear explanations of what data is collected, how it's used, how long it's retained, and whether it's shared with third parties.
- Choose tools that let users delete their data, provide strong encryption, limit permissions, and avoid sharing personal information without consent.

On the other hand, be cautious of apps that offer no privacy settings, use vague language around data use, or have a history of security breaches. Evaluating these factors ensures that prompt crafting remains effective without exposing users or subjects to unnecessary risk.

## 2. Bias Recognition

AI systems can replicate and amplify existing societal biases found in their training data, leading to skewed outcomes. For example, a resume-filtering AI trained on historical company data may prioritize candidates who share characteristics with previous hires, unintention-ally excluding qualified individuals from underrepresented groups. Bias directly affects decision-making, resulting in unfair recommenda-tions, lost opportunities, and even reputational harm for organizations relying on flawed outputs.

To identify bias, regularly audit the tools' outputs, looking for patterns that suggest favoritism or exclusion. There are also open-source tools like IBM's AIF360 or Microsoft's Fair learn you can use evaluate fair-ness of some AI tools. The best defense against bias is the knowledge or awareness of the possibility bias exists.

### 3. Transparency Assessment

Transparency allows users to understand how and why an AI system generates the responses (outputs) that they generate. Trustworthy AI tools:

- Explain methods clearly and provide supporting documentation related to how the AI tool generates responses (outputs)
- Provides information about the developers and the training data used in the development of the AI tool
- Allows access to interaction and revision histories of the AI tool

Ask questions like: Who built this tool? What data was it trained on? Can I trace the reasoning behind the AI tools outputs?

Red flags include poor documentation, lack of feedback mechanisms, and refusal to explain how decisions are made. Without transparency, it's difficult to correct errors or prevent harm. Understanding an AI's inner workings also helps users refine prompts more effectively.

### 4. Personal Accountability

Even the most advanced AI cannot replace human judgment. Users are responsible for verifying claims, double-checking facts, and reviewing recommendations before acting on them. This includes:

- Cross-referencing sources
- Checking citations
- Seeking corroborating evidence
- Using fact-checking tools like Snopes or Google's Fact Check Explorer

For example, a doctor receiving AI-generated medical advice must validate it against clinical standards. Similarly, financial analysts should confirm AI insights before making investment decisions. When users blindly trust AI outputs, the consequences can include misinformation, economic loss, or physical harm. Maintaining ongoing skepticism and critical thinking will help users ensure that every AI-generated suggestion is filtered through a lens of responsibility.

**Tying It All Together**

Responsible AI use is deeply connected to the principles of good prompt crafting:

- Privacy keeps interactions secure
- Bias checks ensure fairness and inclusivity
- Transparency fosters deeper understanding and control
- Accountability maintains quality and integrity

By practicing vigilance in these areas, users can transform well-designed prompts into trustworthy, impactful, and safe AI-powered solutions.

## COMMON PITFALLS AND HOW TO AVOID THEM

Avoid the Blind Trust Trap by never accepting AI outputs at face value. Relying solely on AI-generated content or decisions can lead to serious mistakes, especially when these systems make errors or provide incomplete information. For instance, there have been cases where medical AI diagnostics produced inaccurate results, leading doctors to misdiagnose patients. In another example from education, students who used AI-based essay scoring tools without double-checking their work sometimes received unfairly low grades because the system failed to properly interpret creative writing styles. To prevent this, always apply human judgment:

✔ Double-check AI results against trusted resources, just as you would review a calculator's answer if it seems off.

✔ Ask critical questions such as: Does the output align with known facts? Are all relevant factors considered?

✓ Regularly practice verification by intentionally searching for errors in AI-generated content. Success here is measurable. Track how often you identify discrepancies before acting, and aim to reduce reliance on unchecked outputs over time.

Establish a collaborative review process for major business decisions involving AI. Use a checklist to ensure every AI-influenced decision is reviewed thoroughly. A sample checklist is provided at the end of this chapter. Customize it to fit your needs and monitor outcomes by tracking decisions that were refined by human input versus those accepted without question.

## AVOIDING RISKS THROUGH REGULAR UPDATES

Ignoring Updates and Terms is a pitfall that can expose your business to security risks or legal complications. When AI providers change their algorithms, privacy policies, or service terms, old habits can quickly become outdated, causing trouble down the line. For example, a company using an AI-powered chatbot might find themselves in violation of new data handling rules if they haven't reviewed recent policy updates. In healthcare, changes to machine learning models could alter diagnosis logic, leaving users unaware of significant shifts in accuracy or reliability. To avoid these issues:

- Create a periodic review schedule, such as monthly checks or reviews following major update notifications.
- Maintain a change log for each AI tool, documenting new features, updated terms, and policy changes.
- Measure success by tracking incidents related to missed updates and aiming for full compliance.

## THE VALUE OF PEER REVIEW IN AI-AIDED WORKFLOWS

Peer review is vital for catching overlooked issues and improving the quality of AI-supported outcomes. In research, multiple scientists should validate AI-generated hypotheses before publication. In corporate environments, teams can review AI-driven reports together, identifying gaps that require human scrutiny. In schools, combining peer review with AI feedback yields more balanced evaluations.

To make peer review effective:

- Establish roles for each session (reviewer, note-taker, decision-maker).
- Use structured comparison checklists, where each participant verifies results independently before reaching a consensus.
- Track effectiveness by monitoring error rates and noting improvements from collaborative reviews.

Consistent peer validation helps teams reduce mistakes and boosts confidence in outcomes.

## BUILDING A CULTURE OF THOUGHTFUL AI USE

Start by taking individual responsibility in questioning AI, then expand into contextual awareness, proactive updates, and team-based oversight. Together, these habits create a resilient framework for responsible AI use.

The focus should always remain practical: combine machine efficiency with critical human insight. Monitor your progress and refine your approach continuously to ensure AI enhances your decisions rather than undermining them.

## ETHICAL USE OF AI: A LONG-TERM COMMITMENT

As AI technology evolves, ethical use will remain a challenge. Privacy, fairness, and transparency must form the foundation of how we interact with these tools. Responsible use goes beyond writing effective prompts—it involves thoughtful data sharing, recognizing potential biases, and understanding how AI decisions are made.

Key habits include:

- Fact-checking outputs
- Asking critical questions
- Thinking independently about implications

Avoid common pitfalls such as blind trust, ignoring context, skipping updates, or bypassing peer review. Keep human judgment central to all decisions. When you blend innovation with insight, you unlock the full power of AI while upholding safety, fairness, and trust.

# Collaborative AI Decision Review Checklist for the Workplace

## 1. Define the Decision Scope

- ☐ Clearly state the decision to be made.
- ☐ Identify if the decision is critical, high-impact, or strategic.
- ☐ Confirm that AI is being used to generate, recommend, or support the decision.

## 2. Assemble the Review Team

- ☐ Include relevant stakeholders (technical, managerial, ethical, and operational).
- ☐ Designate a lead reviewer or decision facilitator.
- ☐ Ensure diverse perspectives are represented (e.g., product, legal, compliance, UX).

## 3. Review the AI Input & Method

- ☐ Verify the data source(s) used by the AI.
- ☐ Confirm that inputs are current, relevant, and unbiased.
- ☐ Check the logic or algorithm used to arrive at the decision.
- ☐ Document the confidence level and known limitations of the AI output.

## 4. Validate Decision Alignment

- ☐ Does the recommendation align with company/team goals?
- ☐ Are there any potential ethical, legal, or customer impact concerns?
- ☐ Have alternative scenarios or viewpoints been considered?

## 5. Discuss with the Team

☐ Share the AI's recommendation and supporting data transparently.
☐ Facilitate a structured team discussion (e.g., pros/cons, SWOT, red team/blue team).
☐ Invite feedback, questions, and challenges to the recommendation.

## 6. Document the Review

☐ Summarize team feedback and final decision rationale.
☐ Record who was involved in the review and their roles.
☐ Note any dissenting opinions or reservations.

## 7. Plan for Follow-Up

☐ Set a timeline to evaluate the outcome of the decision.
☐ Assign responsibility for monitoring and metrics tracking.
☐ Plan a retrospective or review loop to refine future AI decision processes.

## 8. Compliance & Accountability

☐ Ensure compliance with internal policies and external regulations.
☐ Store records in a secure, auditable format.
☐ Review whether the decision process is repeatable and fair.

# CHAPTER 5: THE CHANGING WORLD OF WORK -HOW AI IS RESHAPING CAREERS

LAST YEAR, Evan found himself preparing for yet another round of job applications. He had spent a decade as an accountant, confident in his skills and experience. But this time it felt different. Every company seemed to want candidates familiar with new software or data analysis tools he'd never even heard of. At one interview, the panel asked if he'd ever used AI to detect fraud patterns. Evan had to admit he hadn't—and he walked out feeling uncertain about where he fit in anymore.

Experiences like Evan's are becoming more common. People everywhere are noticing changes in their workplace, sometimes exciting, sometimes confusing. Conversations now swirl around unfamiliar terms like *"AI fluency"* or *"prompt engineering."* New job titles are popping up that didn't exist just a few years ago. While some feel uneasy, others see chances to learn, grow, and take on roles they never imagined before.

# THE NEW JOB LANDSCAPE: ROLES AI IS CREATING

Building on foundational AI concepts covered in Chapter 2, where machine learning and neural networks were explained, these core technologies now shape the way people work. They turn complex data into useful insights at a speed and scale never seen before. This leads to major changes in how jobs are performed across many industries.

In healthcare, AI-enhanced professions are becoming standard. Radiologists use AI-powered software to analyze medical images. This technology pinpoints tumors or abnormalities that might escape human eyes, allowing for earlier, more accurate diagnosis. Pharmacists, too, rely on AI to track medication interactions and predict adverse effects based on massive datasets from electronic records. In hospitals, nurses use scheduling tools powered by AI to create better shift plans and reduce burnout. These examples show how, instead of replacing humans, AI works alongside professionals and amplifies their ability to deliver care efficiently and safely.

In finance, AI tools automate repetitive tasks while enhancing decision-making. Financial analysts now use machine learning to analyze global news and trading trends in real time. In banking, chatbots and virtual assistants handle routine customer inquiries, freeing human staff to solve more complex issues. Accountants increasingly use AI for fraud detection, as algorithms quickly identify irregularities in financial data. These innovations improve productivity and allow professionals to focus on higher-level thinking.

Marketing teams depend on AI-driven analytics to understand consumer behavior. For example, AI reviews social media comments, customer clicks, and purchase histories, to recommend which products should be advertised to which groups. Content creators use AI to draft emails or social media posts, then refine them based on performance data. Marketers also deploy chatbots on company websites to answer questions and collect customer feedback around the clock, improving both engagement and satisfaction. The key is not replacement but augmentation. AI gives workers more tools and frees them from manual analysis so they can focus on strategy and creativity.

Beyond enhancing existing roles, AI creates entirely new careers that did not exist just a few years ago. One example is the AI ethicist. These professionals guide companies to build and use artificial intelligence responsibly. Their duties include drafting ethical guidelines, monitoring for bias, and ensuring transparency in algorithms. An AI ethicist can earn anywhere from $95,000 to $180,000 annually, depending on experience and organization size. Another fast-growing role is the data curator, who is responsible for collecting, labeling, and safeguarding the giant amounts of information AI systems need to function. Data curators are now hired in sectors like retail, healthcare, and logistics. Their salaries range between $60,000 and $110,000 per year, with demand rising as more firms invest in large-scale digital transformation.

Other newly created jobs include prompt engineers, people who design effective instructions for AI models like ChatGPT, and AI trainers who teach software how to interpret language, images, or audio properly. Every sector, from education to manufacturing, hires these specialists to help fine-tune AI for specific business needs. Market studies reveal that demand for such roles has jumped more than sixfold since 2018, and companies actively seek talent willing to grow their capabilities in this space.

To succeed in today's workforce, professionals must reach a level of AI fluency. This means understanding how AI tools operate and how to apply them to daily work, even without knowing how to code. AI fluency includes using platforms such as Copilot, ChatGPT, or DALL-E to organize data, simplify workflow, or brainstorm ideas. Many professionals master prompt engineering, crafting clear requests that get the best output from generative AI. Even in customer service or HR, employees utilize automated scheduling, resume review tools, and natural language processing chatbots. Coding remains valuable for some, but non-technical AI adoption is widespread and accessible to most occupations. More than half of hiring managers say they will not hire someone lacking basic AI literacy, yet only one in 500 job listings specifically requires coding knowledge.

## PREPARING FOR CHANGE

In the age of artificial intelligence, adaptability often begins with a growth mindset. Consider a marketing manager whose company recently adopted an AI-powered campaign analytics platform. Rather than resisting the change, she saw it as an opportunity to develop new skills. She actively sought feedback from her manager and colleagues, identified gaps in her knowledge, and kept track of every error message or difficulty she encountered. By scheduling weekly one-on-one sessions with a data specialist, she actively sought constructive criticism and gradually improved her proficiency, making her a more valuable employee.

Examples like hers are increasingly common. On platforms like LinkedIn, you'll find stories of graphic designers evolving into prompt engineers for generative AI tools, and customer service representatives transitioning into roles where they train AI systems. These examples prove that seeking input, welcoming mistakes as learning moments, and setting clear goals for enhancing skills and abilities make all the difference.

Celebrating small wins reinforces a growth mindset. After using an AI-based scheduling app to save hours each week, a team leader shared her experience in a staff meeting, inviting colleagues to exchange questions and tips. This kind of openness not only boosts individual confidence but also nurtures a workplace culture that values adaptation. Such openness not only boosts individual confidence but also helps build a culture where adaptation is valued. Keeping a feedback journal, requesting specific advice after project reviews, and participating in group brainstorming sessions are practical behaviors that foster this mindset. Each positive reinforcement builds resilience and transforms challenges into stepping-stones.

Staying aware of industry trends is another essential part of preparing for change. Professionals can subscribe to key newsletters such as "AI Weekly" or "MIT Technology Review." Joining communities like AI-focused Slack groups or attending webinars through resources like Microsoft's AI Business School gives early access to new developments. Setting aside fifteen minutes each Monday to scan top tech blogs or joining monthly virtual meetups makes trend monitoring manageable. Following major companies—like Amazon's Machine Learning University or Google's "AI for Everyone" initiative—on social media platforms delivers bite-sized updates straight to your feed.

Proactive trend awareness leads to smarter career decisions. A sales analyst who noticed an increase in automated reporting tools positioned herself for a promotion by enrolling in a short online course about AI data visualization before her peers did. Early awareness allows individuals to anticipate shifts, identify required skills, and pivot their efforts before change becomes mandatory. A weekly habit might include reading three articles from trusted sources, saving actionable insights, and discussing them in team huddles. Over time, these actions create a sharper sense of where industries are heading and how to prepare.

At the heart of all successful adaptation strategies lies continuous learning. Rather than cramming knowledge under pressure, professionals benefit more from consistent, scheduled learning routines. Try setting aside two evenings each week for structured learning completing courses like Coursera's Introduction to Artificial Intelligence or participating in IBM's "New Collar" Jobs Program. Balance self-paced study with interactive activities such as problem-solving workshops or emotional intelligence training areas where humans continue to shine alongside AI. Rotate between video tutorials, online discussion forums, and hands-on projects to reinforce learning. Set quarterly goals, such as mastering a new tool or giving a presentation on recent AI advancement, to track your progress.

Developing a growth mindset in the age of AI requires a strong foundation in ethical principles such as those discussed in Chapter 4 that guide responsible technology use. Using AI responsibly goes beyond technical proficiency; it involves understanding critical issues like privacy, data security, and algorithmic bias. Awareness of these factors helps users steer clear of common ethical missteps. Responsible professionals stay informed by reviewing AI ethics guidelines, participating in company-led discussions on dilemmas, and recognizing when to pause implementation until clearer standards emerge. This ethical grounding not only builds personal credibility but also informs wiser decisions about which AI-related skills to pursue.

Building confidence in technology begins with structured, low-pressure practice. Start with simple tutorials, such as using a free AI-powered summarizer or setting up automated calendar suggestions. Aim to try one new tool per week, documenting outcomes and reflecting on what made the process easier or more difficult. From there, experiment with small projects like customizing chatbot responses or exploring no-code data analysis tools. Share your experiences with a mentor or peer group to reinforce learning. As your comfort grows, gradually tackle more advanced tools, such as experimenting with machine learning models or joining guided hackathons. Celebrate every milestone completing a tutorial, automating a task, or pitching an AI-enhanced idea as proof of growing competence. These small wins compound over time, transforming hesitant beginners into confident innovators.

AI is not only transforming existing jobs but also creating entirely new career paths. Roles like AI ethicist and data curator are becoming more prevalent, while traditional positions increasingly incorporate smart technologies. This evolving landscape underscores the importance of adaptability. Staying current with AI trends, learning how to use new tools, and understanding chatbot workflows are all essential to remaining competitive.

Ultimately, thriving in this fast-changing environment depends on being flexible and committed to lifelong learning. It's not about knowing everything at once, but about maintaining a willingness to explore, learn from failure, and make informed choices especially where ethical use of AI is concerned. Developing habits like seeking feedback, celebrating progress, and staying curious creates a solid base for lasting success. In a workplace shaped by constant innovation, curiosity and ethical integrity are among the most valuable traits a professional can have.

# SECTION TWO: FUNCTIONAL APPLICATION

# CHAPTER 6: AI WITHOUT CODING HOW NON-TECHIES CAN SUCCEED

AI ONCE FELT like the exclusive playground of programmers and tech enthusiasts. For a long time, one did need to know how to code or had to understand complex IT systems to tap into AI's potential. The truth is that's no longer the case. Today, friendly digital tools are unlocking new ways for all sorts of people—teachers, event planners, small business owners, even parents at home—to use AI to work smarter and stay organized without ever touching a line of code.

This chapter will guide you through getting started with these accessible AI tools, even if you've never written a script or installed special software before. We'll explore platforms built for beginners, simple onboarding steps, and practical ways to customize your own AI-powered projects. By the end, you'll see that using artificial intelligence is not only possible but also surprisingly straightforward, regardless of your technical background.

## NO CODE, NO PROBLEM: USER-FRIENDLY AI TOOLS

Building on the foundation laid in Chapter 1 about AI's role in everyday life, this section introduces tools specifically designed for non-technical users to leverage AI with ease. Many modern platforms use clear, intuitive designs and helpful guides every step of the way.

Point-and-click platforms lead the charge in making advanced technology simple. Take Softr as an example. Its drag-and-drop interface means you can build a web app or internal tool by arranging blocks like headers, forms, and galleries on your screen. You do not need experience with coding languages or complex setup. With Softr, adding login systems, payment processing, and user permissions becomes as easy as clicking and dragging the right element where you want it. Glide offers another approach for building mobile apps; users pick a template, link up their data via Google Sheets, and then visually arrange features—say, a sales dashboard—with just a few clicks. These platforms let people assemble workflows by connecting pre-made components, such as linking a form submission directly to a notification or an email response. In business settings, someone might automate collecting client information and tracking sales leads by joining these visual elements together. On the personal side, a parent could quickly put together a family event tracker or a budgeting tool. Platforms like Bubble take things a step further, offering power and customization while maintaining no-code interfaces.

Customizable templates offer the next level of accessibility for newcomers. Most leading platforms provide a library of ready-to-use designs tailored to common needs. For instance, someone starting a small business might open Softr or Glide, find a client portal or online shop template, and immediately see sections for products, orders, and messages already in place. The real ease comes in adaptation. Changing colors, swapping images, editing form questions, or adding extra steps takes only a handful of clicks. Suppose a schoolteacher wants to create a quiz for students. By selecting a quiz template on Campaignware, they can update questions, set scoring rules, and add branding to match their classroom theme—all without worrying about code underneath the surface. Even more specialized areas benefit. Dataiku provides templates for building data dashboards or sentiment analysis tools, letting users start with proven layouts and modify them to fit their audience, such as customizing a report to focus on survey responses from a specific department.

Plug-and-play integrations make familiar software even smarter with built-in AI. Many people use Microsoft Word or Slack every day, and recent improvements have brought automated intelligence straight into these environments. In Word, for example, users now take advantage of AI-powered suggestions for clear writing or grammar fixes that pop up automatically. With a click, writers can rephrase sentences or organize reports better. Slack, on the other hand, enables teams to set up smart chatbots using tools like Botsonic, which plugs into the platform without any extra programming. A business owner could activate a bot to answer frequently asked questions or route files between channels, saving time for everyone. When people connect Zapier Interfaces to their workflows, they tap into AI orchestration—letting them automate repetitive work across multiple apps, such as copying customer info from one database to a spreadsheet and sending follow-up emails. Even advanced analytics can be added through platforms like Alteryx, which helps users analyze sales numbers or marketing results through simple, guided steps.

Support resources are the backbone for those learning to use these no-code AI tools. Every top platform offers help centers filled with getting-started articles, instructional videos, and interactive walkthroughs. When a first-time user signs up for Bubble, the recommended tutorial guides them to build a basic CRM app, complete with contact forms and dashboards. If confusion appears, detailed documentation explains each feature, and active forums let users ask questions and find solutions posted by others. Video lessons on YouTube show how to troubleshoot issues or unlock hidden features, all in plain language. Dataiku's collaborative functions mean that several people can work on a data project at once, helping each other within the tool itself. Users often face hurdles like managing their data connections or adjusting app permissions. Articles and quick tips demonstrate common fixes, such as reconnecting a spreadsheet or revising who can view certain content. This community-centered support, combined with accessible explanations and frequent updates, empowers anyone —regardless of background—to solve problems and steadily gain confidence as they explore new possibilities with no-code AI tools.

## PRACTICAL ONBOARDING FOR AI TOOLS

Choosing the right AI tool starts with clarity about what you want to accomplish. If your aim is to organize daily tasks, look for tools that help streamline lists and reminders, such as AI-powered to-do list apps. When evaluating options, prioritize ease of use—does the interface feel welcoming and straightforward? Seek out tools with step-by-step tutorials, clearly labeled menus, and a clean dashboard layout. Learning resources, such as video guides or searchable FAQs, are essential for beginners. Consider community support too; active forums or chat groups make it easier to find answers quickly and troubleshoot roadblocks. For example, Userpilot offers onboarding features that let newcomers experiment with different templates and settings without coding knowledge, while Chameleon and Product Fruits have ready-made walkthroughs, checklists, and pop-ups managed by AI assistants to guide you at every click. Matching the tool with your goals means closely reading the feature list—if automated reminders and smart sorting are important, select an app that highlights these qualities up front.

After selection comes the account setup process. Most modern AI tools keep registration simple. Usually, you start by entering your email address on the home page and choosing a password. Some services also allow sign-up with a Google or Apple account for convenience. After confirming your email, you're directed to a welcome screen that may ask about your main objective—organizing daily tasks, building habits, or managing projects. Answering these questions helps the tool suggest relevant templates or default settings. You might see options to set notification preferences, choose interface themes (like dark mode or high contrast), and import existing data from calendars or spreadsheets. If you get stuck, most onboarding flows now include live chatbots or help modals that offer instant tips or connect you to support. For instance, Freshchat bots can guide you through each decision, clarifying choices about privacy settings or data integration. It's not uncommon to encounter issues like verification emails landing in spam folders or confusion over required fields; in these cases, look for "Resend" or "Need Help?" links beneath the form. Descriptions next to input boxes often explain what information is needed, and screenshots in the help center can visually walk you through each step. Setup screens typically display progress bars so you know how many steps remain before you're ready to use the tool fully.

For your first project, starting small makes success more likely. Let's say you pick an AI-powered to-do list app like those featured in current onboarding platforms. Begin with template selection - a task board, daily planner, or simple checklist are common starters. Click *"Use Template"* to populate your workspace with sample tasks. Edit each item to match your actual goals, such as *"Send client email," "Buy groceries,"* or *"Review project files."* The app's AI might propose deadlines or suggest grouping similar tasks together to increase efficiency. Features like drag-and-drop reordering or auto-categorization make organizing easy. As you add real events, try adjusting tags and priorities to see how the tool schedules or color-codes them. Many tools automatically analyze your past task completion patterns and recommend optimal times for future reminders, boosting your productivity each day. If anything feels unclear, hover over info icons or open a chatbot window for quick guidance. Common problems, such as duplicate entries or lost tasks, are usually resolved by refreshing the page or using built-in undo buttons. Progress indicators show which tips or setup items you've already covered, helping you track your learning.

To build comfort and confidence with your new AI tool, hands-on exploration is critical. Start with practical exercises: create two separate to-do lists for home and work, then enable smart suggestions to compare how task recommendations change. Experiment by toggling notification frequency, changing between list and board views, or adding a recurring weekly chore. Define a metric for success—such as reducing overdue tasks by 30% within your first week. Measure this by viewing the app's built-in analytics panel, which charts completed vs. pending items over time. Every setting adjustment—altering reminder times, switching template types, or customizing colors—affects how easily you spot urgent responsibilities and react to changes in your schedule. These experiments reveal which features best suit your style, whether you thrive with visual boards or concise lists. Continued practice, coupled with reviewing your progress statistics, sharpens your understanding of how AI augments your routines and pinpoints where further customization delivers results.

Mastering these basics lays the groundwork for advanced strategies. As you become comfortable with these onboarding steps, you'll soon learn how to craft effective prompts to communicate more precisely with AI tools, a skill we'll delve into in Chapter 10.

Just keep in mind, anyone can get started with AI using platforms that don't require programming. Whether you're building a simple to-do list or experimenting with creating your own app, today's no-code tools are designed to put powerful technology right at your fingertips. Learning how to use easy-to-use templates, point-and-click interfaces, and plug-and-play features make it possible to customize these tools for your needs. With plenty of supportive resources and active communities ready to help, there's guidance available every step of the way—even if you've never worked with AI before. Keep exploring and trying new features as you go; hands-on practice is the best way to discover what works for you. The journey into accessible AI is all about making things easier, one step at a time. Below you'll find a list of some resources for more in-depth AI learning to explore.

**20 Top AI Learning Tools for Beginners (2025)**

**1. Google AI (Learn with Google AI)**

- Type: Educational Platform
- Best for: Intro to machine learning and AI ethics
- Features: Free courses like 'Machine Learning Crash Course' with real datasets

**2. Microsoft Learn for AI**

- Type: Guided Learning Paths
- Best for: Azure AI tools, cognitive services
- Features: Beginner-level modules with sandbox labs

**3. OpenAI Learn / ChatGPT Playground**

- Type: Interactive AI Experimentation
- Best for: Learning prompt engineering and API usage
- Features: Play with ChatGPT, DALL·E, and Codex; try prompt design

**4. Elements of AI**

- Type: Free Online Course
- Best for: Absolute beginners
- Features: Explains core AI concepts simply (University of Helsinki)

**5. AI For Anyone (AIFA)**

- Type: Non-profit learning hub
- Best for: Teens and adult beginners
- Features: Workshops, gamified learning, practical scenarios

## 6. DeepLearning.AI

- Type: Online Courses (Coursera-hosted)
- Best for: Foundational skills and practical AI building
- Features: AI for Everyone by Andrew Ng

## 7. Fast.ai

- Type: Open source courses
- Best for: Learning AI through hands-on projects
- Features: Code-light intro with pre-trained models

## 8. Teachable Machine by Google

- Type: No-code AI tool
- Best for: Training simple AI models (image, sound, pose)
- Features: Instant training in your browser, visual feedback

## 9. IBM Watsonx / IBM SkillsBuild

- Type: Free educational platform
- Best for: Applied AI use cases, responsible AI
- Features: AI ethics, chatbot building, NLP tools

## 10. Kaggle Learn

- Type: Micro-courses and hands-on notebooks
- Best for: Practicing data science and AI
- Features: Beginner-friendly courses + datasets

## 11. YouTube Channels (Simplilearn, Codebasics, Edureka)

- Type: Video learning
- Best for: Step-by-step tutorials
- Features: Visual walkthroughs of tools like ChatGPT, Bard, and Python ML

### 12. MIT Introduction to Deep Learning

- Type: University Course (Free)
- Best for: Conceptual and visual learners
- Features: Modern deep learning explained in simple terms

### 13. Khan Academy – Intro to AI

- Type: Video lessons and explanations
- Best for: Students and educators
- Features: AI basics through real-life examples

### 14. Buildspace

- Type: Project-based AI learning community
- Best for: Shipping real AI apps
- Features: Guided project challenges + community support

### 15. Replit + Ghostwriter

- Type: Coding playground with AI assistant
- Best for: AI-assisted code writing
- Features: Learn to code while the AI suggests improvements

### 16. Notion AI Playground

- Type: Embedded AI tool
- Best for: Writing with AI and prompt testing
- Features: Great for testing how AI responds to context

### 17. Runway ML

- Type: Visual AI tool
- Best for: Creators and visual learners
- Features: Play with generative AI models visually

## 18. DataCamp – AI Fundamentals

- Type: Online platform
- Best for: Structured AI learning
- Features: AI and ML tracks with interactive coding

## 19. AI Dungeon

- Type: Interactive game
- Best for: Learning how AI language models 'think'
- Features: Sandbox to understand natural language generation

## 20. LearnPrompting.org

- Type: Open-source prompt engineering guide
- Best for: Practicing AI prompt writing
- Features: Gamified prompt challenges and practical examples

# CHAPTER 7: PRACTICAL AI IN THE WORKPLACE

THE REALITY IS that AI is now quietly transforming how everyday tasks get done even in the smallest businesses or most ordinary offices. From scheduling meetings to sorting through cluttered inboxes, tasks that used to eat up hours each week are now handled in seconds. Suddenly, those repetitive chores aren't weighing you down—they're just running in the background, thanks to tools that anyone can use.

Earning extra income has always seemed like a hustle reserved for people who have special skills or insider knowledge. But the rise of AI is changing this picture fast. Now, with just a laptop and an internet connection, everyday people are using smart apps to spin up creative gigs, tap into emerging trends, or simply find flexible work on their own terms. The line between day job and side project has never been more blurred, and it's easier than ever to jump in, even if you're not a tech pro.

# ENHANCING WORKPLACE PRODUCTIVITY WITH INTELLIGENT AI TOOLS

Building on our exploration of user-friendly AI tools in Chapter 6, let's now see how these tools can be applied to boost workplace productivity. Many companies today rely on no-code AI platforms with easy onboarding, which means even team members without technical backgrounds can set up and benefit from powerful automation features. As more organizations move toward accessible technology, the practical uses of AI at work become clear.

Automated scheduling tools are one of the most noticeable ways AI saves time. Instead of juggling countless emails or looking for open slots, AI calendar assistants review your entire schedule, analyze preferred meeting times, and check others' availability. By integrating directly with systems like Google or Microsoft Outlook, these tools flag potential conflicts and suggest windows that fit everyone's needs. If new tasks or emergencies arise, the AI automatically finds a better slot and sends an update to all participants, so you never have to reorganize manually. When teams use these assistants, they often save nearly 45 minutes per day—the time usually lost to back-and-forth coordination. Features like automatic reminders and meeting buffers help people focus on important conversations instead of worrying about logistics. For example, a busy sales manager can set preferences for sales calls, interviews, and internal check-ins, then trust the system to protect blocks of focused work time. Even when last-minute changes come up, Scheduler AI and similar tools handle rescheduling and confirmations, ensuring nothing falls through the cracks.

AI-driven email management has changed the way professionals deal with overflowing inboxes. Smart email solutions analyze incoming messages, sort them by urgency and topic, and flag anything that needs immediate attention. High-priority client requests or urgent project updates land in a dedicated folder, while routine newsletters or notifications are grouped elsewhere. Some AI-powered apps automatically draft responses, reminding you to follow up with key contacts or confirming appointments. This targeted sorting reduces the hours often spent checking and organizing emails, freeing up at least four hours every week. A marketing executive, for instance, might receive hundreds of emails daily. With AI triage features built into their platform, they see only what matters most—critical approvals or last-minute campaign updates—without missing deadlines or losing important details. These tools keep communication flowing smoothly and ensure quick reactions to urgent needs.

Writing professional content is another area where AI offers huge productivity gains. AI writing assistants improve reports, emails, and presentations by offering real-time grammar and style suggestions, so each document sounds polished and clear. They can generate first drafts for proposals, blog posts, or customer service replies, allowing users to simply review and refine. This is especially valuable during tight deadlines when speed is essential. Rather than starting from scratch every time, team members input topics and receive structured text to edit and personalize. Legal teams, marketers, HR reps—all benefit from having instant support as they prepare contracts, press releases, or onboarding instructions. The result is fewer errors, more confidence in written work, and faster turnaround on important communications. AI writing tools also store templates and respond to feedback, getting better at matching a specific tone or message over time.

Routine, repetitive tasks once eaten up much of the average worker's day. Today, AI automates many of these chores, freeing employees to focus on bigger projects. Platforms equipped with machine learning take care of data entry and basic reporting - scanning spreadsheets, importing info into company databases, or cleaning up inconsistent formats. In finance, automated invoice processing tools extract payment details from PDFs and enter them straight into accounting systems, reducing mistakes and speeding up month-end reviews. Sales and operations staff find value in AI bots that send out meeting follow-up notes or pull the latest sales trends from different systems for quick analysis. By tackling tasks like scheduling, updating records, or generating simple charts, AI cuts down on human error and ensures that teams always work with accurate information. For example, a project manager who once spent mornings updating progress trackers can now let an AI agent gather updates and show them in real-time dashboards. What took hours now takes minutes, boosting not just efficiency but team morale as well.

These workplace tools are designed for fast setup and everyday usability, making advanced automation available to everyone. Employees gain extra hours each week by using AI scheduling, smart email sorting, writing aids, and task automation. The less time spent on routine work, the more energy teams can invest in creative thinking, problem-solving, and building relationships. As organizations adopt smarter tools, productivity rises—not just for individuals but for entire departments, helping businesses stay responsive and competitive in a changing world.

## UNLOCKING NEW INCOME STREAMS: AI FOR SIDE HUSTLES

As we explored in Chapter 5, AI is reshaping careers and creating new income avenues. This section delves into concrete ways to leverage AI for side hustles and freelance work. Anyone, even those without a technical background, can start earning using AI-powered tools and micro-tasking platforms. Here are four practical methods you can start today.

AI-powered content creation tools open the door for people interested in writing, graphic design, and creative services. Tools like **Jasp**er (formerly Jarvis) and Copy.ai make it easy to create blog posts, product descriptions, or ad copy by entering a few keywords or prompts. These platforms generate drafts within seconds, saving hours of manual writing. Canva's Magic Write and DALL-E introduce straightforward methods for visual content—users can produce images, social media graphics, or even unique artwork just by describing what they want. Designers no longer need advanced software skills; AI streamlines each step from ideation to final edit. With these solutions, beginners can offer social media packages or branded content to small businesses. Income can range from $50 to $500+ per month as a part-time creator, with some experienced freelancers charging $25–$100 per project or more, especially when combining writing and visuals. AI brainstorming features—like Notion's "AI Assist" or ChatGPT itself—

help jumpstart projects by generating lists of business names, content themes, or marketing strategies on command. For someone growing a personal brand or supporting clients, this speeds up idea development and broadens their service menu.

AI also transforms how people access market research. Several platforms now provide automated market analysis, such as TrendHunter or Exploding Topics. With these tools, users see which products, topics, or services are gaining popularity online, supported by real-time data. You don't have to be a data scientist; you simply log in and review the results—often displayed in easy-to-read charts or reports. These AI-driven insights show where demand is rising. For example, Etsy sellers use tools like eRank to monitor trending crafts, while bloggers rely on Ubersuggest to find profitable keywords for articles or affiliate sites. By following actionable recommendations provided by these platforms, newcomers have launched stores or blogs that quickly gained traction and sales, sometimes reaching $200–$1,000 in extra monthly income based on the niche selected. AI turns vast market information into simple steps, helping anyone make informed decisions before investing time or money.

Micro-tasking opportunities are another proven way to earn through AI. Platforms like Appen, Remotasks, and Amazon Mechanical Turk let users contribute directly to AI system improvement. Tasks include image labeling, video tagging, content moderation, or training chatbots to respond naturally. Most roles do not require special technical skills, attention to detail and basic communication are enough. Depending on the complexity, entry-level compensation ranges from $3 to $10 per hour for very basic tasks, but specialized roles such as AI chatbot trainers or community managers can reach $15–$30 per hour if you develop experience. This work model is highly flexible; workers select assignments when they have free time, fitting them around other jobs, school, or family commitments. Many people use micro-tasking as a supplement to regular employment or as an introduction to the gig economy. Real stories show that consistent micro-taskers can regularly boost their monthly budget by $100–$600, depending on dedication and chosen platform.

For those offering freelancing or teaching services, AI dramatically expands the possible offerings. Freelancers can combine Grammarly with Jasper to polish and enhance client documents at speed. Marketers deliver smarter campaigns by analyzing customer messages using natural language processing add-ons like MonkeyLearn. Teachers and tutors use tools such as Quillionz or Canva to make personalized worksheets, quizzes, and lesson plans in minutes. This approach allows educators to take on more students or sell ready-made resources online. On Upwork and Fiverr, listing "AI-boosted" writing or design gigs lets providers stand out and charge higher rates —sometimes $25–$100 per hour for unique, fast-turnaround projects. In addition, professionals developing skills in prompt engineering— writing instructions for AIs like ChatGPT—have seen new high-paying gigs emerge at $25–$100 per hour, often without needing computer science backgrounds.

All these methods connect through the same goal: using AI to break down barriers and unlock new streams of income. By choosing the right tool or platform, setting up profiles on freelancing websites, and consistently applying for gigs, anyone can tap into the expanding world of AI-powered side hustles. Whether it's creating content, researching markets, micro-tasking, or upgrading freelance offers, these practical steps turn AI advancements into everyday earning opportunities.

From automating schedules to sorting emails or drafting content, these tools save time on repetitive tasks, giving you more room to focus on creative projects and building better connections with your team or clients. It's not just about getting through your to-do list faster. It's also about feeling less overwhelmed by the little things that used to take up hours each week. Real AI solutions can make daily work easier and open up new ways to earn money.

AI-powered platforms simplify earning extra income, even if you don't have a technical background. Whether you want to start freelancing, try micro-tasking, or use creative tools to launch a side hustle, practical steps are within reach. The move toward user-friendly AI means everyone has a chance to boost productivity and unlock new possibilities. By staying open to learning and trying out different options, you can use AI to shape work that fits your skills and lifestyle. Below you'll find a list of some of the top performing AI Platforms to explore.

# 20 TOP AI CONTENT CREATION & VIRTUAL ASSISTANT PLATFORMS (2025):

**Text & Copywriting Tools**

## 1. Jasper AI

- Best for: Marketing copy, blog posts, social media content
- Highlights: Pre-built templates, tone control, team collaboration features

## 2. Copy.ai

- Best for: Quick marketing content and product descriptions
- Highlights: Budget-friendly, user-friendly, supports multiple languages

## 3. Write Sonic

- Best for: SEO blog posts, ad copy, chatbot responses
- Highlights: Built-in AI article writer, including Chat Sonic assistant

## 4. Sudowrite

- Best for: Fiction writers and creative storytelling
- Highlights: Tools for story arcs, character development, and tone guidance

## 5. INK AI

- Best for: SEO-optimized writing
- Highlights: Real-time SEO scoring and content optimization

**Image & Graphic Generation**

### 6. Canva AI (Magic Studio)

- Best for: Visual design, presentations, image editing
- Highlights: Features like Magic Write, Magic Edit, and Magic Design

### 7. Adobe Firefly

- Best for: AI-powered design and photo editing
- Highlights: Part of the Adobe suite, supports commercial use

### 8. DALL·E (OpenAI)

- Best for: Creating illustrations and photo-realistic images from text
- Highlights: Integrated into ChatGPT, generates visuals via prompts

### 9. Night Cafe Studio

- Best for: Artistic and printable artworks
- Highlights: Style transfer, batch creation, and active community

### 10. Craiyon

- Best for: Quick idea visualization
- Highlights: Simple interface, free to use

**Video & Audio Creation Tools**

**11. Runway ML**

- Best for: Text-to-video generation and video editing
- Highlights: Gen-3 Alpha, real-time editing, background removal

**12. Synthesia**

- Best for: AI avatar videos and training content
- Highlights: 140+ avatars, supports over 120 languages

**13. Pictory AI**

- Best for: Converting blogs or scripts into videos
- Highlights: Voiceovers, captions, automatic summarization

**14. Lumen5**

- Best for: Short-form videos for marketing and social media
- Highlights: Drag-and-drop editor, AI storyboard generation

**15. Murf AI**

- Best for: Realistic AI voiceovers
- Highlights: 120+ voice options, emotional tone control, pitch/speed editing

## Multimodal & All-in-One AI Studios

### 16. Descript

- Best for: Podcast and video editing
- Highlights: Edit videos via text, AI voice generation

### 17. Genmo

- Best for: Creating videos, 3D, and images
- Highlights: Collaborative AI agents that respond to prompts

### 18. Eleven Labs

- Best for: AI voice synthesis and dubbing
- Highlights: Emotion-driven speech, multilingual support, voice cloning

### 19. Notion AI

- Best for: Productivity, content planning, note-taking
- Highlights: Embedded in Notion, flexible AI prompt tools

### 20. Claude Artifacts

- Best for: Collaborative and interactive content creation
- Highlights: Create live, editable documents directly from AI chats

# 20 TOP VIRTUAL ASSISTANT AI PLATFORMS (2025):

## 1. ChatGPT (OpenAI)

- Best for: Personal and business virtual assistance
- Key Features: Custom GPTs, voice/chat modes, coding, writing, brainstorming, scheduling

## 2. Microsoft Copilot (365)

- Best for: Enhancing productivity in Word, Excel, Outlook
- Key Features: Integrated into Microsoft Office, summarization, presentation design

## 3. Google Duet AI (Workspace)

- Best for: Gmail, Docs, Sheets, and Meet
- Key Features: Smart replies, meeting scheduling, doc summarization

## 4. Siri (Apple)

- Best for: Device control and voice-activated tasks on Apple devices
- Key Features: Voice commands, scheduling, smart home integration

## 5. Amazon Alexa

- Best for: Smart home management and voice interactions
- Key Features: Skill ecosystem, routines, reminders, audio playback control

## 6. Replika AI

- Best for: Mental wellness and companionship
- Key Features: Empathetic chat, voice support, coaching, journaling

## 7. Ada

- Best for: Automated customer support (enterprise)
- Key Features: Multilingual AI chat, self-service workflows, integrations with Zendesk and Shopify

## 8. Amelia (IPsoft)

- Best for: Enterprise service automation
- Key Features: Natural interaction, emotion-aware responses, voice and chat support

## 9. x.ai / Scheduler AI

- Best for: Automated scheduling and meeting coordination
- Key Features: Email-based scheduling, smart availability tracking

## 10. SoundHound / Houndify

- Best for: Voice interfaces and real-time AI responses
- Key Features: Speech recognition, custom wake words, IoT compatibility

## 11. My AI (Snapchat)

- Best for: Conversational support and casual Q&A
- Key Features: ChatGPT integration, visual interaction

## 12. Notion AI

- Best for: Task management, writing, and productivity
- Key Features: Built-in summarization, brainstorming, to-do generation

## 13. Grammarly GO

- Best for: Enhanced writing and editing
- Key Features: Rewrites, tone adjustments, ideation, email creation

## 14. Tidio AI

- Best for: Customer service for small businesses and e-commerce
- Key Features: AI chatbots, canned replies, Shopify integration

## 15. Youper

- Best for: Mental health tracking and journaling
- Key Features: CBT-based AI conversations, mood insights, daily mental health check-ins

## 16. Krisp AI Assistant

- Best for: Enhancing remote meetings
- Key Features: Background noise removal, real-time notetaking, meeting summaries

## 17. Kore.ai

- Best for: Customer and employee engagement
- Key Features: Omnichannel bots, text and voice support, domain-specific intelligence

### 18. Paradox (Olivia)

- Best for: HR and recruiting automation
- Key Features: Resume screening, scheduling interviews, FAQ chat

### 19. Magic (GetMagic.com)

- Best for: Personal task outsourcing
- Key Features: AI + human hybrid assistant, 24/7 task handling

### 20. Heyday by Hootsuite

- Best for: Customer support on social platforms
- Key Features: AI-powered conversations for Instagram, Messenger, WhatsApp

# CHAPTER 8: AI FOR PERSONAL GROWTH AND SELF-IMPROVEMENT

"REMIND me to call Mom after work," I mumbled, still half-asleep as I shuffled into the kitchen. My phone lit up with a chirpy response, promising to set the reminder and even suggesting a time based on my calendar. It felt strange at first, letting my device nudge me about personal things. But soon enough, these gentle prompts became background music in my daily routine. Little by little, my phone started knowing what I'd forget before I did. It wasn't magic. It was technology quietly learning who I am and how I live.

Learning something new used to mean finding a class or poring over a textbook late at night. Now, I can pick up Spanish while brushing my teeth or get instant feedback on my latest attempt at coding, all thanks to clever apps that seem to know exactly where I struggle and where I shine. It's like having a private tutor who never gets tired of explaining things differently until I finally get it. The more I use them, the more they adjust, encouraging me to keep going, even when I feel stuck.

# AI INTEGRATION INTO DAILY LIFE: ENHANCING PERSONAL ROUTINES AND EFFICIENCY

Building on the workplace efficiency tools covered in Chapter 7, such as automated scheduling and AI writing assistants, these personal applications show how AI's benefits extend beyond work into everyday life. Just as AI streamlines office tasks, it now enhances our personal routines, creating a seamless flow between professional and personal productivity.

Personal assistant apps are often the heart of this transformation. Apps like Google Assistant, Siri, and Alexa help people manage their time with automatic reminders, smart suggestions, and real-time calendar updates. For example, once you start using a digital assistant to schedule appointments or set reminders, it quickly learns your habits - like when you usually wake up or travel to work. Over time, it can suggest optimal alarm times, anticipate meetings, and provide friendly nudges for things that need attention based on past behavior. These assistants sync with calendars, email accounts, and even to-do lists, offering a unified organizational system. If you confirm a meeting by email, your AI assistant can add it to your calendar and send a reminder an hour before. Integrated routines allow users to manage not just work tasks but also grocery lists, family events, and personal goals from the same app, significantly reducing mental clutter.

This integration goes further when AI connects with other devices used daily. Voice-operated features in assistant apps make them accessible for vision-impaired or mobility-challenged users who can use spoken commands for everything from checking the weather to sending texts. These assistants tie together smartphones, tablets, computers, and even car infotainment systems, ensuring users can access personal information or perform actions wherever they are. Such capabilities offer new levels of independence and convenience, blending naturally into routines without extra effort.

Entertainment and shopping have also become smarter thanks to AI-powered recommendations. Streaming platforms like Netflix and Spotify do more than store your favorite songs and shows. They analyze your viewing history, search patterns, and even pause moments to create highly personalized suggestions. For instance, Netflix uses what you watch, how long you watch it, and which movies you skip to choose the next set of recommendations. Spotify's Discover Weekly playlist relies on your listening habits, liked songs, and playlists, then combines this with data from users who share similar interests to deliver a fresh batch of tracks each week. Online stores follow suit. Amazon looks at your browsing history, previous purchases, and items you clicked but didn't buy. This enables special deals, reminders for restocking household goods, and suggestions for future buys that match your taste. These recommendation engines turn overwhelming choices into easy decisions, making shopping and entertainment smooth and enjoyable.

Wellness tracking through AI is another area where technology blends effortlessly into personal life. Wearable devices like Fitbit, Apple Watch, and Garmin track health metrics such as step count, sleep cycles, and heart rate around the clock. These gadgets don't simply measure progress; AI-driven insights translate raw data into helpful suggestions. If the tracker sees a drop in your activity or an unusual heart pattern, it might prompt you to move or check in with a health professional. Over weeks and months, these systems observe trends to offer tailored advice, like recommending earlier bedtimes or reminding you to hydrate during busy days. Some advanced assistants are developing emotional intelligence, allowing them to pick up on mood shifts and offer support if they sense rising stress or anxiety.

Home automation brings AI's touch to every corner of a modern living space. Smart thermostats such as Nest and Ecobee adjust temperatures on their own after learning when you're typically home or away. Voice-activated lighting systems like Philips Hue can change colors, dim lights, or turn off lamps with simple phrases. Security cameras and smart doorbells record and notify homeowners about visitors or unexpected movement, while locks and alarms can activate at scheduled times or when directed by voice. Refrigerators, ovens, and washing machines respond to commands or carry out tasks automatically, freeing up time and energy for other activities. AI assistants coordinate all these devices into routines. At night, a single "Goodnight" command can lock doors, lower lights, and set the alarm, providing peace of mind and comfort.

In all these examples, AI takes the stress out of decision-making and routine management. By quietly collecting information about user behaviors, preferences, and needs, these technologies personalize and automate tasks that once took much more time. The result is a daily experience where chores fade into the background, freeing individuals to focus on the things they care about most. Whether juggling a busy family calendar, searching for something fun to watch, monitoring one's health, or securing a home, AI personal assistants have become invaluable companions, always learning, always adapting, and always ready to make life a little easier.

## LIFELONG LEARNING POWERED BY AI

Adaptive learning platforms are reshaping how individuals pursue lifelong education. Unlike traditional static courses, modern adaptive systems such as DreamBox and Smart Sparrow harness artificial intelligence to analyze a learner's responses and performance at every step. When a user solves a math problem or works through a reading comprehension task, the system instantly records their choices, time spent, and accuracy. The next lesson is then dynamically adjusted in difficulty and focus, making tasks harder if the learner excels or offering targeted review for any mistakes. This approach ensures that every learner progresses at an optimal pace, maximizing retention and minimizing frustration. For instance, in K-12 classrooms, platforms like McGraw-Hill ALEKS use knowledge space theory and AI algorithms to continuously re-evaluate student mastery, ensuring gaps are filled before advancing.

Moving into language and skill acquisition, AI-powered apps like Duolingo and Babbel have become everyday companions for millions seeking new proficiencies. These tools utilize advanced speech recognition technologies to assess pronunciation in real time. When a learner practices speaking a phrase, the app's AI evaluates tone, accent, and correctness, delivering precise feedback almost instantaneously. Gamification lies at the heart of these experiences: streak counters, badges for new milestones, and leaderboards create engaging incentives. Duolingo's personalized learning paths, for example, rearrange skills and topics based on an individual's strengths and recurring errors. If someone struggles with verb conjugations but breezes through vocabulary, subsequent lessons introduce more targeted practice on verbs while keeping engagement high through reward loops. In professional upskilling, Coursera and LinkedIn Learning employ similar adaptive tracks powered by AI to recommend modules most relevant to current career goals.

AI also plays a crucial role in feedback and improvement systems across educational tools. Platforms like Gradescope leverage machine learning to evaluate open-ended homework submissions, highlighting strengths and flagging common errors. In coding education, GitHub Copilot provides real-time suggestions and reviews code for best practices as developers work. Progress visualization features, often seen in Khan Academy and Lumosity, act as powerful motivators: users receive charts and graphs that display their development over time. These visual dashboards do more than track completion. They highlight areas needing improvement, but they also highlight growth, providing learners with positive visual reinforcement that encourages consistent advancement. Recommendations for additional exercises, video lessons, or peer challenges, stem directly from the AI's ongoing assessment, enabling focused and meaningful practice sessions without overwhelming the user with irrelevant material.

Learning doesn't happen in isolation, and AI-driven community connections now enhance collaborative growth. Modern platforms integrate intelligent matching algorithms that pair learners with suitable study partners or mentors based on proficiency level, interests, and even preferred communication styles. Coursera's discussion boards and Discord servers use these algorithms to suggest group projects or peer reviewers whose experiences closely align with the user, fostering productive exchanges. AI also identifies supplementary resources by analyzing patterns in the learner's questions, quiz results, and browsing habits. On Skillshare, after completing a course segment, the platform often recommends specific project prompts, articles, or expert videos tailored to the challenges a learner has faced. This extends beyond basic content. AI considers subtle factors like learning speed, recent feedback, and chosen specialization areas. By embedding these features, platforms encourage networking and knowledge sharing, creating vibrant ecosystems where motivation and success become collective.

Looking ahead, the skills developed through these AI learning tools will prove invaluable when we explore prompt crafting essentials in Chapter 10. The adaptive learning techniques discussed here lay the groundwork for mastering AI communication skills. Whether fine-tuning grammar in real time, visualizing personal progress, or connecting with fellow learners globally, AI's integration into lifelong education establishes a seamless path from basic knowledge acquisition to advanced digital literacy.

AI is quietly weaving itself into our daily routines and the way we learn new things. From smart assistants managing our schedules to streaming services that get to know our tastes, AI has found thoughtful ways to lift the weight of everyday decisions from our shoulders. These technologies are helping us bring order to busy lives, making simple tasks even simpler, and opening doors for more personal time and well-being.

When it comes to lifelong learning, AI offers a fresh path for growth and self-improvement. Tools adapt to our needs, offer real-time feedback, and connect us with others who share our goals. This journey isn't just about keeping up—it's about being empowered to chart our own course with support right when we need it. The subtle presence of AI helps us focus on what matters most: developing skills, building confidence, and enjoying the learning process along the way.

# CHAPTER 9: REAL-WORLD SUCCESS STORIES: AI IN ACTION

WHAT IF THE skills you need to get ahead are no longer about working harder, but about knowing how to use artificial intelligence as your personal advantage? The old rules of success—long hours, strict routines, and doing everything yourself—are being rewritten every day. Today, real breakthroughs come to those who let AI handle the repetitive work, make smarter choices, and turn challenges into quick wins.

Ever wonder how AI actually fits into the busy lives of real people or what it looks like when technology works quietly in the background, making things smoother? This chapter shares stories from all kinds of people—professionals in offices and freelancers at home, teachers in classrooms, and parents juggling family schedules. You'll see exactly how AI helps everyday tasks feel less overwhelming, letting people focus on what matters most to them.

# WORKPLACE WINS: PROFESSIONALS THRIVING WITH AI

As we explored in Chapter 5, AI tools can boost productivity and open new income streams. Building on that foundation, here we showcase how professionals across fields are thriving by integrating AI into their workflows. Let's look at some inspiring real-world cases.

A marketing director at a consumer electronics brand faces the challenge of interpreting massive amounts of customer data to guide advertising campaigns. Traditionally, the team would spend weeks sorting surveys, social media posts, and sales numbers, sometimes missing critical market shifts. By implementing Google Analytics and IBM Watson Marketing, the director transforms the workflow entirely. The team now runs predictive analytics to segment audiences and identify emerging trends instantly. During a recent product launch, an AI-powered dashboard flagged a sudden uptick in social interest for a feature they hadn't promoted heavily. The marketing manager quickly adjusted digital ads to highlight this feature, resulting in a 22% higher click-through rate than previous launches. Post-campaign analysis showed a 19% lift in sales conversions and a 30% reduction in ad spend waste. These results underscore how real-time AI analytics empower marketers to pivot instantly and capture more value from every campaign.

In a busy primary school, a fourth-grade teacher juggles the diverse needs of students who progress at different rates. Previously, creating custom lesson plans or adapting assessments for each learner took hours after school. By bringing in AI-powered learning games such as DreamBox and Squirrel AI, the educator personalizes assignments automatically. The software evaluates student responses in real time and adapts difficulty, pacing, and hint delivery accordingly. For example, one student struggles with fractions but excels at geometry; the AI directs her coursework to focus on fractions until mastery. Over a semester, standardized test scores show a 15% improvement in math proficiency across the class, with struggling students catching up faster than before. Parents report greater engagement, citing excitement to "beat the AI game" at home. The teacher regains valuable planning time and sees improved learning outcomes, discovering that intelligent content adaptation levels the playing field and fosters growth for all learners.

A freelance copywriter confronts the daily grind of answering client emails, scheduling calls, and chasing payments—tasks that erode precious creative hours. To reclaim lost time, he integrates Chatfuel and ManyChat AI chatbots into his business website and project management tools like Trello's Butler AI. Routine queries about pricing and availability receive instant, consistent responses. Automated invoicing via FreshBooks AI reduces late payment follow-up to a single weekly review. In the first month alone, the freelancer saves ten hours per week—time redirected toward developing new writing products and pitching high-caliber clients. As a direct result, output of long-form articles doubles within two months, and monthly earnings jump by 35%. Clients appreciate the fast response and clear communication, leading to more repeat business. The freelancer realizes that automation doesn't diminish quality; it amplifies his creative reach while maintaining a personalized client experience.

A small specialty food store owner faces inventory headaches when popular items sell out unexpectedly or seasonal promotions leave shelves overstocked. Data entry and tracking in spreadsheets produce frequent errors and slow reactions to demand changes. Introducing ShelfSense AI and Shopify's predictive analytics modules lets the owner monitor POS data and customer behaviors in real time. When the system notices increased searches for gluten-free snacks, it suggests new purchase orders and flags which products should be displayed more prominently. Acting on these suggestions, the owner reorders bestsellers and rotates displays based on AI insights. Within a quarter, spoilage drops by 40%, and overall sales grow by 17%. Inventory turnover stabilizes, and the entrepreneur reports less stress and more focus on developing new products. These gains confirm that even small businesses can implement smart technology to optimize decision-making, demonstrating the vast potential AI holds for daily operations no matter the company size.

Professionals in marketing, education, freelancing, and small business ownership are succeeding by matching their unique challenges with targeted AI solutions. These stories show how AI tools help individuals adapt, free up time, and drive measurable improvements in performance. While these examples center on workplace wins, the seeds are planted for AI to play a similar role in daily life beyond professional settings, suggesting even larger transformations ahead.

## AI IN EVERYDAY LIFE: FITNESS, FINANCES, SHOPPING, AND TIME MANAGEMENT

Sarah had always found it hard to stick with exercise routines. Busy workdays and unpredictable schedules made keeping active feel impossible. One evening, frustrated after missing yet another workout, she tried an AI-powered fitness app recommended by a friend. On her first use, the app asked for just 10 minutes of her time and a few questions about her goals and current fitness level. Instead of following generic videos, Sarah received weekly plans designed specifically for her schedule and needs. Exercises adjusted in real time based on her performance; when knee pain flared, low-impact alternatives appeared instantly. Motivational nudges celebrated every improvement—big or small—which helped Sarah look forward to each session instead of dreading it. Over three months, workout consistency soared from once a week to five times per week, increasing her activity rate by more than 80%. As the app recognized her progress, adaptive challenges introduced new exercises, preventing boredom and sustaining her motivation. These personalized feedback loops, powered by AI's ability to analyze patterns and respond immediately, turned what was once a struggle into a sustainable habit.

While Sarah improved her wellness, Alex faced his biggest challenge in managing money. Before using AI budgeting tools, most of his paychecks seemed to vanish without a clear reason, leaving him anxious at the end of each month. Integrating his bank accounts with an AI-driven finance app revealed automatic expense categorization that sorted hundreds of transactions into clear categories—food, utilities, entertainment, and more—within seconds. For the first time, Alex saw that dining out consumed nearly 25% of his monthly income. The app didn't stop there; it offered targeted recommendations, like setting a weekly dining-out budget and identifying cheaper supermarkets nearby. With proactive alerts and tips, Alex started meal prepping, reducing food expenses by $120 per month. AI reminders notified him two days before bills were due, helping him avoid late fees—a problem he'd faced twice last quarter. After six months, he'd built an emergency savings cushion of $1,800, exceeding his original goal thanks to AI-driven nudges that suggested incremental savings as his habits improved.

Maria's experience with smart shopping showed how AI can help maximize value from every purchase. For years, she bought items during weekend sales, often missing better deals online or forgetting to claim available discounts. When Maria began using an AI-enabled shopping assistant, the system continuously monitored prices on everything from electronics to shoes. One example stood out: she saved $130 on a new vacuum cleaner after receiving a timely price drop alert. Personalized product recommendations soon replaced scrolling through endless listings. When she needed a winter coat, the app factored in her past purchases, preferred colors, and sizing details to present three options within her set range. Decision-making became quicker and more confident. For grocery runs, AI compared prices across local stores and flagged a 20% discount at one branch of her favorite market. On average, Maria reported saving about $60 monthly, all while cutting shopping time by half. The satisfaction went beyond discounts—she appreciated knowing she was making informed choices without the exhaustion of comparison shopping every week.

Efficiency extended into time management for Tom, a working parent managing two kids' school schedules, meals, and family activities. Mornings used to be chaotic attempts to coordinate soccer, music lessons, and project deadlines with his partner's appointments. Once Tom adopted an AI-integrated calendar, life began to change. This digital tool synced seamlessly with his email, mobile device, and shared family accounts. It automatically detected overlapping commitments and suggested optimal times for activities such as grocery runs or doctor visits. Shopping lists generated through the app matched upcoming meals with store promotions, saving both money and decision fatigue. In one instance, the calendar rearranged his son's dentist visit and daughter's recital to avoid conflicts, while also finding a slot for a much-needed family walk in the park. Stress levels dropped noticeably; Tom estimated that planning family tasks now took only 15 minutes per week, down from an hour previously. Over the course of three months, missed appointments dropped to zero and family dinners increased from two nights to five per week, resulting in better communication and stronger connections at home.

Whether fostering fitness motivation, reducing financial anxiety, streamlining shopping, or simplifying complex schedules, these stories highlight AI's power to enhance daily living across multiple fronts. Through personalization, predictive insights, and automation, AI not only saves time and money but also helps people achieve their personal and family goals with less effort and more confidence. These everyday applications of AI set the stage for our next chapter, where we'll dive deeper into mastering AI prompts to get the best results from your AI tools.

These stories show just how much AI can make a difference in both work and daily routines. By leaning into smart tools, people are finding more time for what matters, making better decisions, and staying ahead of challenges that once felt overwhelming. Whether it's a teacher unlocking every student's potential, a copywriter freeing up hours for creativity, or a store owner keeping shelves stocked just right, the positive changes are clear. These examples show that when AI is used thoughtfully, it doesn't replace the human touch—it makes it stronger.

In everyday life, AI supports healthier habits, smarter spending, and easier schedules. It notices things we might miss and helps us act on them before they become problems. People aren't just saving money or boosting productivity—they're also lowering stress and feeling more confident in their choices. As technology keeps growing, so will the helpful ways it can fit into our lives. The real takeaway is that with the right approach, AI can be a reliable partner in reaching our goals, big or small.

# SECTION THREE: PRACTICE

# CHAPTER 10: MASTERING AI PROMPTS: GETTING THE BEST RESULTS FROM AI TOOLS

AI prompt engineering is the practice of crafting clear, specific, and well-structured inputs (prompts) to guide an AI system toward generating useful and high-quality outputs. Picture the first time you tried using an AI assistant for either work or personal tasks. Perhaps you gave simple requests like "write a letter" or "analyze these numbers," only to realize the results didn't quite fit your needs. It's easy to assume the technology isn't ready or that there's something wrong with your request. Many people wonder if there's a trick to getting clearer replies—whether it's for writing an email, summarizing a report, or brainstorming new ideas. How do some users get the exact information they need while others are left frustrated?

As you experiment a bit more, you will start to see patterns—certain small changes in how you ask questions can lead to much better answers. This chapter is about tracing those steps and understanding how small adjustments to your prompts can turn average interactions into surprisingly helpful ones.

# FUNDAMENTALS OF PROMPT CRAFTING

In order to make AI work well for you, you need to communicate clearly and be specific. When you give the AI a good prompt, you're setting it up for success. This was shown in Chapter 9 with stories of people who clearly communicated their needs and got better results faster. This led to less frustration and higher quality work. Let's dive into the basics of effective prompts, which include being clear about your goals, providing context, and refining your prompts as needed.

Providing clear goals or defining your objectives is about knowing what you want your end result, or output, to be before you ask for it. Vague requests can leave AI guessing, resulting in broad or irrelevant answers. Instead, thinking through your objective and turning it into a clear instruction for the AI to follow ensures better results. Imagine you need to draft an email asking for project updates. A weak prompt might be, "Write an email to my team." This leaves the AI with too many possibilities, such as topic, tone, or recipients. Breaking the request down further helps to guide the AI.

For example, you are a project manager needing an update from your team.

**Original weak prompt**: "Write an email to my team."

◆ **Improvement process:**
First, specify the purpose: provide project update.
Next, mention the action required: reporting their latest progress.
Third, add a deadline.

> ➤ **Enhanced prompt:** "Write a polite email to my project team asking them to send their current progress on the marketing campaign by Friday."

Why does this work? By setting a clear goal, you direct the AI to create an email that matches your intent, includes all necessary details, and sets expectations. Consider applying this method to data analysis too:

rather than saying, "Analyze this spreadsheet," clarify what trends or figures interest you most.

Providing enough context is the next ingredient. The more background you give, the smarter the AI's response will be. If you only say, "Summarize this report," the output may be too generic. Adding context like your industry or intended audience sharpens the result. For example, stating, "Summarize this quarterly sales report for our company's board of directors. Focus on revenue growth and main challenges," gives the AI important direction. This means you spend less time rewriting or clarifying later, since the AI understands the who, what, and why behind your request.

Let's practice with another scenario. Example: You want AI to write social media posts about a new product.

**Weak prompt**: "Write posts about our new product."

◆ **Step-by-step improvement:** Explain your brand personality, target audience, and desired outcome.

> ➤**Enhanced prompt:** "Create three engaging Facebook posts announcing our eco-friendly reusable water bottle. Make sure the language is cheerful, mention the product's sustainability benefits, and encourage customers to share their own tips for reducing plastic waste."

Why does this work? These details provide context which allows the AI to tailor its messaging to your goals and audience, which saves you editorial effort later on.

**Context can include:**

✔ Format: how should AI structure the response – list, table, essay, etc.
✔ Style or Tone: how should AI communicate the response – conversationally, humorously, in an academic manner, etc.
✔ Examples: offer AI specific examples of how you would like the response. You can tell AI "Use the same template or format

as...." Or "Write this for a _____audience (professional, female over age 40, high school-age male, etc.).

Iterative refinement is the process of repeating a process with the goal of improvement. In the case of Prompt Crafting, it means transforming a good prompt into a great one. This means repeating this prompt improvement process until AI provides you with exactly the output you're looking for! Your first attempt is rarely perfect, so it's important to experiment, reflect, and adjust. Start with your original prompt, record the AI's response, note any issues, revise your wording, and document the improved outcome.

For example, if you asked, *"Explain climate change simply,"* but found the answer confusing, you could revise it to: *"Explain climate change to middle school students using everyday examples and short sentences."* Compare the responses to see if the revision produced a clearer and more age-appropriate explanation. Through this process, you will learn what information guides the AI most effectively.

A practical tip is to notice when your prompt produces vague, off-topic, or overly brief answers. This signals a need for more detail or context. Remember, finding the right approach comes with a mix of patience and curiosity. Each iteration brings you closer to the results you want, no matter your profession or task. Whether you're developing creative stories, drafting business emails, or organizing to-do lists, these strategies apply across the board. Creating a simple Prompt Writing Journal can be very helpful. A template to create a Prompt Writing Journal can be found at the end of this chapter.

**To pull everything together, try this exercise**.

**Scenario:** You run a small consulting business and need an email sent to a client recapping a completed project and requesting feedback. Begin with a basic prompt: "Write an email to my client about our recent project."

After reviewing the AI's draft, identify missing details: *Which client? What project?*

What feedback do you want? Revise the prompt: *"Write a professional email to Sarah Kim, thanking her for partnering on the market research project we just finished. Include a summary of key milestones, express appreciation for her input, and ask for feedback on our team's communication and final report."*

Compare both drafts side by side. Notice how the enhanced prompt delivers a message that feels personal, comprehensive, and actionable.

Document your steps, responses, and takeaways in your prompt journal to build your skills even further. Mastering these basics means you'll be fully prepared to tackle advanced techniques like instruction layering and output measurement, turning every AI interaction into a productive collaboration.

## REFINING PROMPTS FOR PRECISION AND IMPACT

Layering instructions is a powerful way to transform a basic prompt into one that produces exceptional results. When you break down complex tasks into clear, step-by-step directions, AI models can process the request more effectively and deliver more accurate responses. For example, suppose you want help drafting a marketing email. Instead of simply asking, *"Write a marketing email for our new product,"* you can scaffold your request using layers: begin by specifying the target audience, then describe the product's top benefits, followed by the tone you want, and finally ask for a call-to-action at the end. This layered approach ensures each element receives focused attention and reduces ambiguity, yielding outputs tailored to your goals.

**A typical sequence might look like:**

1. Identify the reader,
2. Highlight the main features,
3. Incorporate persuasive language, and
4. End with a strong call-to-action.

Common pitfalls include being too vague at any step or overwhelming the model with too many details in a single instruction. Success looks

**97**

like receiving an output that touches on every specified point in a logical order.

Once you receive an AI-generated output, analyzing it carefully is key to improvement. Start by reading the response with fresh eyes, looking for places where information may be missing, off-topic, or unclear. If a section feels weak—perhaps the benefits aren't compelling, or the call-to-action lacks punch—pinpoint what was lacking. To adjust, revise your prompt by adding details or clarifying the intended style. For instance, if your first attempt at a marketing email seems generic, refine your prompt to say, *"Emphasize why this product saves customers time and use energetic language."* Re-run the improved prompt, and you'll likely discover the next reply fits your expectations more closely.

Before-and-after examples make this process concrete: a weak prompt like *"Summarize this article"* can be improved to *"Summarize the main arguments and conclusions from this article in three sentences, suitable for a newsletter."* The revised version directs the AI to trim irrelevant material and sharpen the summary. Don't hesitate to iterate multiple times, each time focusing your feedback on specific areas for growth. Watch for improvements not just in accuracy, but also in how well the output matches your unique requirements.

To take your refinement skills even further, adopt a practical framework for evaluating the quality of AI responses. Develop a list of criteria—such as relevance, accuracy, and completeness—and score each output from 1 to 5 on these aspects. Relevance means the answer sticks to your original question; accuracy checks the facts or logic; completeness confirms nothing important is left out. For example, a strong research summary will directly address the prompt, use correct data, and avoid omitting crucial findings. In contrast, a weak response might ramble or leave gaps. Setting up a simple scoring grid lets you track which types of prompts consistently get high marks, and helps you quickly spot patterns over time. If you notice certain requests regularly fall short on completeness, tweak future prompts to be more specific about the end goal. Common missteps include confusing length for quality or ignoring factual mistakes if the writing sounds polished. True success shows up when your average scores rise, reflecting outputs that are both useful and trustworthy.

Keeping a prompt journal makes systematic improvement possible. Documenting each prompt you try, alongside the generated result and what worked or didn't, brings hidden lessons into focus. Set up your journal with categories such as intention (why you wrote the prompt), actual outcome (what the AI produced), adjustments made, and ideas for the next revision. Over time, reviewing these notes uncovers strategies that lead to better results and reveals stumbling blocks you tend to encounter. Even noting failed attempts can be valuable as they highlight exactly where tweaks are needed. To get started, set aside a few minutes after each interaction to jot down your process and observations, either digitally or on paper. This habit turns trial-and-error into intentional skill-building, helping you become more efficient and creative with each session.

Practicing these techniques builds a foundation for responsible and dependable AI use. By learning to craft precise prompts and systematically refine them, you minimize misunderstandings and gain reliable outputs—a crucial factor when decisions or information depend on AI-generated results (Kuka, 2025). Troubleshooting tips for common challenges include double-checking for ambiguous words, breaking down multi-part requests, and not settling until the output genuinely serves your needs. Try exercises like rewriting a recent prompt to add an extra layer of context, or experimenting with your scoring system on different types of replies. Evaluate your progress weekly using your journal entries to spot ongoing opportunities for improvement.

Developing AI skills is about more than just getting quick answers, and developing prompt refinement skills isn't just about getting better answers— it's about making sure the AI is working for you in the best way possible, and about using AI responsibly, especially when your work or decisions have broader impacts. The ability to shape clear, effective prompts puts you in control. You're building the habits needed for making sound judgments and holding AI tools accountable for their outputs. Mastery here prepares you for the vital conversations around ethics, decision-making, and trust in AI interactions that follow.

As you practice setting clear objectives, adding background information, and reflecting on each outcome, you'll notice your interactions becoming smoother and more useful. These next few pages will provide some tools to assist you in developing habits that will lay the groundwork for responsible and thoughtful use of AI, giving you the tools to handle more advanced tasks with confidence.

# How to Write Effective Prompts for AI

## Step 1: Clearly Define Your Objective

Ask yourself: *What do I want the AI to do?*
Be specific: *Are you trying to generate an article, summarize a book, solve a math problem, or design a logo?*
Example (unclear): *"Tell me about marketing."*
Better: *"Summarize 3 effective digital marketing strategies for small businesses."*

## Step 2: Provide Enough Context

Include relevant background details the AI should know.
Mention who the audience is, the tone you want, or any constraints.
Example: *"Write a 300-word product description for a fitness tracker targeted at busy moms in a conversational tone."*

## Step 3: Specify the Format

Tell the AI how to structure the response: list, table, script, essay, checklist, etc.
Example: *"List 5 bullet points highlighting the benefits of meditation for beginners."*

## Step 4: Set the Style or Tone

Ask for a tone: professional, conversational, inspirational, academic, humorous, etc.
Example: *"Write a humorous Instagram caption about coffee on Monday mornings."*

## Step 5: Use Roleplay or Persona-Based Prompts (optional)

Tell the AI to *"act as"* a certain type of expert to improve quality.
Example: *"Act as a financial advisor and explain the difference between an IRA and a 401(k) to a beginner."*

### Step 6: Include Examples or Templates (if needed)

If you want a specific structure or style, give an example for the AI to follow.
Example: *"Use the same format as this: [Insert sample text]. Now rewrite it for a different audience."*

### Step 7: Ask Follow-Up or Refinement Questions

AI works best interactively. Ask for revisions, expansions, or simplifications.
Examples:
- *"Great. Now shorten this to under 100 words."*
- *"Make it more emotionally engaging."*

### Step 8: Test and Tweak

If the first result isn't what you need, revise your prompt:
- Add clarity
- Remove ambiguity
- Break complex tasks into parts
Refinement Example:
Initial: *"Write an article on AI."*
Better: *"Write a 500-word beginner-friendly article on how AI is transforming education, in an informative yet simple tone."*

### Bonus Tips

- Avoid vague terms like *"tell me something"* or *"do it better."*
- Use constraints (*"in 3 sentences," "under 100 words"*) to keep responses focused.
- For creativity tasks (stories, poems, branding), let the AI know how much freedom it has.

———

**AI Prompt Writing Checklist:**

☐ Have I clearly defined what I want the AI to do?
☐ Did I include enough background context?
☐ Did I specify the format I want the output in (list, table, paragraph, etc.)?
☐ Did I mention the desired tone or style?
☐ Have I asked the AI to act as a specific expert or role (if needed)?
☐ Did I provide an example or template for reference?
☐ Have I asked follow-up or clarifying questions to improve the result?
☐ Have I reviewed and tweaked the prompt for clarity and precision?

# AI Prompt Writing Journal

Date: _____

*Prompt*:

_____

_____

AI Response Summary: ___ adequate ___ inadequate

What Worked:

_____

What to Improve:

_____

*Revised Prompt*:

_____

_____

AI Response Summary: ___ adequate ___ inadequate

What Worked:

_____

What to Improve:

_____

*Revised Prompt*:

_____

_____

AI Response Summary: ___ adequate ___ inadequate

**What Worked:**

_____

**What to Improve:**

_____

# CHAPTER 11: LEARNING BY DOING – HANDS-ON AI ACTIVITIES

"I HAD no idea a chatbot could organize my week better than I could," a friend once confessed over coffee, laughing as she recalled her failed attempt at juggling meetings and reminders. Her story isn't unusual, many of us begin our journey with new technology feeling skeptical or even intimidated. But with time, the right tools often turn into surprisingly helpful allies in our daily lives.

Getting comfortable with AI requires curiosity, patience, and a willingness to explore. This chapter invites you to roll up your sleeves and engage with AI directly. It's a practical way to discover what works best for you and what doesn't. Along the way you'll track personal wins, reflect on challenges, and consider how to keep improving.

## BUILDING PRACTICAL AI CONFIDENCE THROUGH GUIDED EXERCISES

With the foundational concepts from earlier chapters in mind, including the lists of tools like ChatGPT, NotebookLM, Perplexity, Manus, Gamma, and OpenAI's 4o Image Generator, you're now ready to move from theory to practice. These familiar applications are more than just digital assistants; they can significantly boost your productivity and efficiency when used with confidence. This section guides you through four practical activities designed to build hands-on experience, boost your confidence with AI, and prepare you for self-assessment in the next chapter.

**Activity 1: Exploring Chatbots for Daily Tasks**

**Objective:** Learn how to interact with conversational AI chatbots to simplify everyday activities and problem-solving.

To begin, access an online chatbot such as ChatGPT or Perplexity. You will need to create an account with these online platforms, but most have a free version so you should not have to spend money to access these AI tools. Once you've created your account, you can begin.

Start by typing a clear, specific question: *"Plan my Thursday schedule around these meetings: 9AM team sync, 1PM project review."* Notice how the AI organizes your day. Try a follow-up command like, *"Add a reminder to call the bank at 3PM."* Observe if the chatbot integrates this new instruction seamlessly. It may respond by asking how you would like to be reminded, or what tools can be used to complete the task.

Expected outcomes include receiving logical suggestions, organized lists, or quick answers. Track your progress by noting how often you get the information you need on the first try. If the responses seem off-topic, try rephrasing your request, for example *"Please summarize the pros and cons of remote work,"* to see if results improve.

*Troubleshooting Tip: If responses seem odd or inaccurate, return to simplified questions, breaking requests into smaller parts, and watching out for typos. Over time, you'll notice that your queries become more effective, and the assistance provided feels natural, helping you save time each day.

## Activity 2: Automating a Routine Task With AI

**Objective:** Set up an AI automation to streamline a repetitive task, such as sorting emails or scheduling meetings.

Choose a task, like organizing incoming emails by priority or flagging calendar invites. Check your email platform for built-in AI features. Microsoft Outlook will prioritize important emails into two tabs: Focused and Other. There are also advanced features in Microsoft 365 to filter emails into high, normal, or low-priority emails that can be sorted on demand. These filters require you to set up some rules to guide AI.

Example: Create a rule to sort all emails from "HR" into your "High-Priority" folder.

Configure the AI to recognize specific keywords or frequent contacts.

Expected outcomes include:

- Fewer missed messages
- Cleaner inbox
- Less time spent manually sorting emails

Measure your success by tracking how much time you save over a week. If some emails don't sort correctly, double-check the rules, keyword filters, or sender settings. Periodically review these settings to ensure they align with your current needs.

Success indicators:

- Processing twice as many emails in half the time,
- Never missing a meeting or reminder.

*Troubleshooting Tip:

- Confirm that AI features are turned on,
- Make sure the app has permission to access your messages.

## PERSONALIZING AI RECOMMENDATIONS EXERCISE

Objective: Tailor AI-driven suggestions in everyday apps to better match your preferences and routines. Whether in music streaming apps, retail platforms, or content aggregators, take time to explore the personalization settings or feedback features. Give input on recommendations such as marking *"I like this artist,"* rating suggested products, or flagging articles as *"relevant"* or *"irrelevant."* Many platforms allow you to hide topics or prioritize specific interests. Each piece of feedback helps the recommendation engine learn more about your tastes, gradually refining the suggestions you receive.

You can measure your progress by reviewing how well playlists, shopping recommendations, or reading lists reflect your actual interests. If suggestions become less relevant, revisit your feedback history and update any mistaken likes or dislikes. These small, consistent adjustments lead to more enjoyable listening, smarter shopping, and more personalized news feeds.

### Structured Observation and Reflection Exercise

**Objective:** Document your AI interactions, track successes and challenges, and reflect on learning progress.

Use a simple template:

1. Date,
2. Tool Used,
3. Task Attempted,
4. Outcome (Success/Challenge),
5. Notes for Improvement.

For example:

1. March 14 –
2. ChatGPT –
3. Generated project summary –
4. Success –

5. Edit prompt for even briefer output next time.

**Edit prompt for even briefer output next time**

Keep this journal over several days. Reflection prompts might include:

- ☐ What tasks became easier?
- ☐ Where did I struggle?
- ☐ Which features were most helpful?
- ☐ Did I notice improvements in speed, accuracy, or convenience?

Every entry provides insights into where you're quickly mastering AI and areas for further exploration.

Progress indicators include increasing ease of using the tools, shorter troubleshooting time, and visible gains in productivity. If you face consistent barriers, consider reviewing earlier chapters or searching for updated tutorials to fill knowledge gaps. As you collect experiences, you'll see the benefits extend beyond individual tasks—building general comfort, curiosity, and skill with AI.

Always prioritize safety during these exercises. Make sure to verify information with trusted sources, refrain from sharing sensitive personal data, and carefully set your privacy options. Retain the authority to make final decisions on significant matters. Using AI responsibly allows you to fully benefit while minimizing potential risks. Together, these practices create a foundation for self-reflection, enabling you to honestly assess your skills and set goals for future growth in the next section.

## REFLECTING ON YOUR AI LEARNING JOURNEY AND PLANNING FOR CONTINUED GROWTH

With the hands-on experience you've gained using different AI tools, take a moment to recognize how far you've come. Navigating new technology isn't always simple, yet with steady practice, you now bring both skill and practical understanding to every interaction. This journey gives you real-world insight that supports responsible choices. As you reflect, consider how your confidence has grown, not just in operation, but also in identifying when and how to use AI responsibly. Also identify any weaknesses where you might lack confidence, so you know where to focus your learning needs.

Start by creating a comprehensive skills inventory. Evaluate your grasp of AI fundamentals—do you understand how machine learning models work and what drives their predictions? Reflect on your comfort using popular AI tools: which interfaces feel intuitive, and where do you still hesitate? Rate your prompt crafting abilities; can you write instructions that produce accurate results, or do you often get unexpected outputs? Examine your ethical awareness: are you aware of biases in AI datasets, and do you strive to choose fair, transparent solutions? Finally, think about your problem-solving capabilities: have you applied AI to address challenges at work, in your studies, or daily life? List each competency, rate your current level, and highlight areas of strength and those needing more attention.

Move to recognizing achievements. What is your most successful AI implementation so far? Recall a time when an AI tool helped you solve a tough problem or speed up repetitive work. Name the biggest challenge you overcame. Maybe it was figuring out how to integrate data sources, breaking through confusion about terminology, or troubleshooting inaccurate outputs. Think about unexpected discoveries; perhaps along the way, you stumbled upon a shortcut or realized a creative way to use AI that wasn't obvious initially. Note the impact these wins have had on your work habits or personal projects. Did AI give you extra time, help you deliver better results, or spark new ideas? Telling your own success stories builds motivation and highlights practical value from your learning journey.

Now identify growth opportunities. Which AI topics remain difficult or confusing? Do you need more practice with prompt writing, troubleshooting errors, or distinguishing trustworthy resources from hype? List knowledge gaps clearly. Ask yourself: Where do I want to be more independent? Which skills would unlock bigger benefits for my job or interests? Pinpointing these specifics will help shape your next steps.

Set clear, realistic goals using a goal-setting framework. For each area needing improvement, define specific targets, such as *"write better prompts for customer support bots"* or *"learn to spot bias in recommendation systems."* Make sure every goal is achievable and measurable. Next, give yourself a timeline. For instance, aim to complete an online tutorial on ethical AI use this month. Identify what you'll need, such as a list of trusted tutorials, relevant communities, or a mentor who can provide feedback.

List practical implementation steps: block study sessions on your calendar, join a peer group, and schedule short check-ins to track your progress. By being intentional, you stay motivated and see real improvement.

Create your personalized development roadmap. First, prioritize short-term enhancements: perhaps mastering common AI tools or getting comfortable with new features. Outline long-term objectives too, like contributing to an open-source AI project, developing your own automation workflow, or deepening your ethical expertise as AI adoption grows in your industry. Gather resources. Bookmark guides from credible sites, connect with active AI discussion groups, consider following leading AI researchers, or even setting up a recurring call with a mentor. Record your progress with regular notes or a shared journal. Celebrate small wins to sustain momentum and update your plan as you reach milestones or as new AI advancements emerge.

Remember, this is an ongoing process. As AI evolves, continual self-reflection will help you pivot and thrive. Each cycle of assessment reveals fresh strengths and uncovers spaces for growth. Embracing this mindset ensures you don't just keep pace, you become adaptable in a landscape that changes quickly. These habits lay the foundation for future-proofing your skills and empower you to lead confidently in a world shaped by AI.

**Reflection**

By taking time to try new tools, automate simple tasks, personalize recommendations, and reflect on your experiences, you are building genuine confidence that goes beyond just understanding theory. These exercises aren't just about learning how the tools work, they're about discovering how you work best alongside them, noticing your strengths, and spotting areas where you'd like to improve.

As you move forward, remember that every attempt, whether it felt smooth or a bit challenging, is another step toward mastering these important skills. Stay curious, keep experimenting, and don't be afraid to ask yourself what's working well and what could use more attention. This mix of action and honest reflection will help guide your ongoing growth, supporting you in reaching both your current learning goals and those you set for the future.

# CHAPTER 12: YOUR AI ROADMAP STRAYING AHEAD FOR LONG-TERM SUCCESS

IN THE FAST-PACED world of artificial intelligence, staying current with the latest developments and continuously improving your skills is essential for long-term success. AI is a dynamic and evolving field, and those who keep learning and adapting will be the ones who thrive. This chapter will guide you through creating a structured plan tailored to your goals that will help you stay ahead in the AI journey.

## MAKE AI LEARNING A DAILY PRACTICE

When it comes to mastering Artificial Intelligence, consistency matters more than intensity. Daily exposure helps reinforce concepts, build fluency, and keep your motivation high. If you want AI to become second nature, it needs to become a part of your everyday life.

## Create a Dedicated AI Learning Schedule

Start by setting aside specific time blocks in your weekly schedule for AI learning. Treat it like an appointment you can't miss—because it's an investment in your future.

- Choose a consistent time each day that fits naturally into your routine. Morning study helps set the tone for the day, while evening sessions can solidify new information before bed.
- Use digital calendars, reminders, or habit-tracking apps like Habitica, Notion, or Google Calendar to stay on track.
- Don't overcommit. Start with just 15–30 minutes a day and increase gradually as you build confidence.

The goal is not to cram, but to create rhythm and reliability.

## Focus on One Concept at a Time

Avoid the temptation to learn everything at once. AI is a vast field, and trying to absorb too much too quickly can lead to frustration and burn out. Instead:

- Break down complex topics into smaller chunks. For example, spend one week focusing solely on understanding machine learning, another on natural language processing, and another on AI tools you can use without coding.
- Use a *"theme of the week"* approach to stay focused and deepen your understanding.

## Rotate Between Learning Modes

Everyone learns differently—so mix up how you engage with material:

- **Read** short articles, tutorials, or ebooks.
- **Watch** explainer videos and online course modules.
- **Listen** to podcasts while exercising or commuting.

- **Do** hands-on practice using AI platforms or building small projects.

Variety keeps your brain stimulated and increases retention by activating different learning pathways.

**Make AI Part of Your Environment**

To keep AI top-of-mind, build passive learning into your day:

- Set your browser homepage to a tech news site or an AI-focused blog.
- Follow AI educators and thought leaders on LinkedIn or X (formerly Twitter).
- Add an AI-themed widget or flashcard app to your phone's home screen for spontaneous learning.

Even a 5-minute scroll through curated content can reinforce concepts and expose you to fresh insights.

## TRACK PROGRESS AND CELEBRATE SMALL WINS

Tracking your learning creates a sense of accomplishment and encourages momentum:

- Keep a learning journal or digital log to record what you've studied, new terms you've learned, and ideas for future exploration.
- Use progress trackers like Trello, Notion, or even a physical notebook to map your milestones.
- Celebrate small wins—completing a tutorial, understanding a concept, or successfully using an AI tool—these moments add up and fuel your confidence.

By making AI learning a daily habit, you're not just acquiring skills, you're creating a mindset that embraces growth, discipline, and long-term success. Over time, these short, consistent learning sessions will build a powerful foundation for navigating the fast-paced world of artificial intelligence.

## Stay in the Loop with Industry Trends

In the world of Artificial Intelligence, things change fast. The first step in staying in the loop is to be aware of emerging trends in AI. New tools emerge, algorithms improve, and ethical debates unfold almost daily. If you're not actively paying attention, it's easy to fall behind. Tracking trends helps you anticipate changes to learn relevant new skills early. But staying current doesn't have to be overwhelming, it just requires a smart, intentional approach to gathering information.

## Curate Your AI News Feed

Start by building a trusted stream of high-quality, relevant content. This means subscribing to curated newsletters that deliver top stories, research breakthroughs, and practical tools straight to your inbox.

Here are a few standout options:

- *The Batch* by DeepLearning.AI – Weekly summaries of the most important developments in AI, written for learners and professionals alike.
- *Import AI* by Jack Clark – Focuses on the implications of cutting-edge AI developments, often connecting technical advances to broader social issues.
- *AI Weekly* – A digest of articles, tools, and research papers, useful for both beginners and advanced practitioners.

These newsletters do the hard work of sifting through the clutter so you can focus on what matters most.

**Listen to Industry Podcasts**

Podcasts are an excellent way to absorb insights during passive moments—like while commuting, walking, or cooking. Many feature interviews with leading experts, hands-on advice, and emerging use cases.

Recommended shows include:

- *Practical AI* – Focuses on real-world applications of AI across industries, often featuring developer tools and business strategies.
- *AI Today* – Hosted by Cognilytica, this podcast explores enterprise AI adoption, trends, and lessons learned.
- *Lex Fridman Podcast* – In-depth conversations with AI pioneers, scientists, and tech leaders (ideal for deep dives).

Even listening to one episode per week can help sharpen your understanding of where the field is heading.

**Follow Thought Leaders on Social Media**

Social platforms like **LinkedIn**, **Twitter/X**, and **Reddit** are treasure troves of real-time updates and community insights. By following influential voices, you get immediate access to breaking news, commentary, and community discussions.

Some tips:

- On **LinkedIn**, follow AI educators, company pages (like OpenAI, Hugging Face, DeepMind), and trending hashtags like #AI, #MachineLearning, and #NoCodeAI.
- On **Reddit**, visit subreddits such as r/MachineLearning, r/Artificial, and r/learnmachinelearning.
- On **X (formerly Twitter)**, keep tabs on research labs, authors of popular AI frameworks, and ethical AI advocates.

Engaging with these platforms even 10–15 minutes a day can keep your awareness sharp and expose you to global conversations.

**Save & Organize Key Resources**

Use bookmarking tools like Pocket, Notion, or Raindrop.io to save articles, videos, and podcasts for later review. You can even create topic-specific folders (e.g., "Prompt Engineering," "AI Ethics," "No-Code Tools") for easy reference and review. This not only reinforces learning but also gives you a personalized library of insights you can revisit any time.

The bottom line is staying in the loop with AI trends doesn't require hours of effort. It just takes a system. Schedule dedicated time into your weekly calendar to:

- Subscribe to and review trusted newsletters.
- Listen to thought-provoking podcasts.
- Follow experts and communities online.
- Organize and revisit what you find.

By making trend-tracking part of your weekly routine, you'll not only stay informed, you'll stay relevant, confident, and ready to seize new opportunities as AI evolves.

## BE READY TO ADAPT

In the ever-evolving landscape of Artificial Intelligence, one truth stands firm: *what works today may be obsolete tomorrow*. New tools are released monthly, models are retrained weekly, and industry standards shift rapidly as technology advances. To not only survive, but thrive, in this environment, you must become adaptable. Adaptability isn't just a skill. It's a mindset, and one of the most valuable assets you can develop in the era of AI.

## Embrace Change with Curiosity, Not Fear

It's natural to feel uncertain when something you just learned gets replaced or improved. But instead of resisting change, lean into it with curiosity. Ask:

- *What can this new tool teach me that I didn't know before?*
- *How can this innovation help me work smarter, not harder?*
- *What skills will make me more effective and efficient in the long run?*

By approaching AI as an ongoing conversation and not a final destination, you free yourself from the pressure to "know it all." You become an explorer, not just a student.

## Reassess and Refresh Regularly

Growth doesn't happen by accident. It's a result of intention. Make it a habit to explore new frameworks, datasets, and platforms that emerge in the community, and experiment hands-on to deepen your understanding. Schedule a quarterly *"skills audit"* to reflect on your progress and realign with current trends:

- Are the tools you're using still relevant and efficient?
- Have better or simpler solutions emerged?
- What are employers, creators, or innovators currently prioritizing?
- Which emerging skills (like prompt engineering, ethical AI, or no-code automation) are gaining traction?

These check-ins help you avoid stagnation and ensure your knowledge stays sharp and applicable.

## LET GO OF WHAT NO LONGER SERVES YOU

It's easy to cling to the first AI tools you became comfortable with. But if you're serious about growth, you must be willing to outgrow your old ways. That might mean:

- Switching from one platform to another that better suits your goals
- Abandoning outdated workflows for more automated or efficient ones
- Upgrading your thinking, not just your tools

Think of it as pruning a tree—cutting away the old makes space for stronger, more fruitful branches.

### Growth Requires Flexibility

Remember: adaptability isn't weakness, it's a superpower! It allows you to pivot with confidence, seize new opportunities, and remain relevant as the world shifts. And in a field like AI, that flexibility can make all the difference between falling behind and leading the way.

By adopting this mindset, you'll become more than just an AI learner, you'll become a lifelong innovator, ready for whatever comes next.

## BUILD YOUR NETWORK TO GROW FASTER

AI might be powered by algorithms, but progress in the field is deeply human. From discovering new tools to solving complex problems, the most valuable insights often come not from solo effort—but from the people you surround yourself with. Whether you're a student, a professional, or a self-taught enthusiast, building your network is one of the fastest ways to level up.

## Why Community Matters in AI

The pace of change in AI can be dizzying. No one can keep up with everything alone. That's where your network comes in. By connecting with others who are also learning, building, or leading in AI, you gain:

- Access to fresh perspectives on tools, use cases, and challenges.
- Exposure to real-world applications you might not discover on your own.
- Motivation and accountability keeps your learning consistent.
- Opportunities for mentorship, collaboration, or even job referrals.

In short, being part of an AI community multiplies your learning and amplifies your results.

## Attend Virtual and In-Person Events

You don't need to be a Silicon Valley insider to access valuable events. There are AI-focused meetups, webinars, and workshops happening around the world, many of them free or low-cost.

- Explore Meetup.com or Eventbrite for AI events in your area.
- Look for virtual conferences and summits, such as NVIDIA's GTC or the DeepLearning.AI community events.
- Attend hackathons, which offer immersive learning experiences, team building, and exposure to real-time problem solving.

Whether you're attending to learn, ask questions, or just listen in, showing up is the first step to building meaningful connections.

## Join Online AI Communities

Not all networking has to be face-to-face. Online communities are thriving, and they're often more inclusive and diverse than local circles.

Start here:

- **Reddit**: Subreddits like r/MachineLearning, r/learnmachinelearning, and r/artificial are excellent places to find answers, read discussions, and discover resources.
- **Discord servers**: Many AI education platforms (like Coursera, Kaggle, or Fast.ai) and creators host active Discord communities where users help each other troubleshoot problems, share links, or collaborate on ideas.
- **LinkedIn groups**: Join professional groups dedicated to AI or machine learning, and follow industry leaders to stay inspired and informed.

## Learn Together to Go Further

Studying with others can deepen your understanding faster than going it alone. When you explain a concept to someone else, you reinforce your own knowledge. When someone else shares their workflow or project, you gain practical insights you might never encounter through tutorials alone.

**Ways to connect:**

- Create or join a virtual study group that meets weekly via Zoom or Google Meet.
- Use platforms like StudyTogether, Notion, or Slack to collaborate on notetaking, assignments, or project development.
- Partner up with a peer to build a real-world project together, even something small like a chatbot, automation workflow, or AI-powered resume analyzer.

These shared experiences will not only grow your skills but also build lasting professional relationships. If you want to grow quickly in AI, don't do it alone. Build your network intentionally by:

- Attending events (virtually or locally),
- Joining active communities on Reddit, Discord, or LinkedIn,
- Participating in study groups or collaborative projects.

The connections you make will expose you to new ideas, unlock resources, and expand your opportunities. AI may be built on data— but your progress will be powered by people. There are multiple tools provided in these last few chapters. The tools in the Daily AI Learning Toolkit can help you create your roadmap to useful AI Learning that will assist you to succeed in this evolving new world.

# Daily AI Learning Toolkit

### ✅ Daily AI Learning Checklist

| DATE | TASK |
|------|------|
| | Set a dedicated for AI learning today (15-30 minutes) |
| | Focus on one concept or skill area |
| | Engaged with one learning format (video, article, podcast, or course) |
| | Practiced with an AI tool or did a mini project |
| | Logged your learning in your AI journal or tracker |
| | Celebrated a small win or milestone |
| | Reviewed previous concepts or notes |
| | Reflected: What did I learn today and how can I use it? |
| | Activities to Stay Current |

### Daily AI Learning Journal Page Template

📑 Date: _____ ⏰ Time Spent Learning: _____ minutes

💭 Focus Topic of the Day

_____

_____

🔍 What I Studied or Practiced

_____

_____

## Key Takeaways

_____

_____

## How I Applied It (or Plan To)

_____

_____

## Small Win or Breakthrough

_____

## Quick Review

What topic or skill would I like to revisit? _____

What's one thing I want to explore next? _____

# ✅ Staying Current with AI Checklist

| Task: | Completed/Comment: |
| --- | --- |
| Subscribe to AI newsletters (e.g., The Batch, Import AI, AI Weekly) | ☐ |
| Read at least one AI newsletter per week | ☐ |
| Listen to one AI podcast episode weekly (e.g., Practical AI, AI Today) | ☐ |
| Follow at least five AI experts on LinkedIn or X (Twitter) | ☐ |
| Join an AI-focused subreddit (e.g., r/MachineLearning, r/learnmachinelearning) | ☐ |
| Engage in at least one online AI discussion or comment thread per week | ☐ |
| Bookmark interesting AI articles using Pocket, Notion, or Raindrop.io | ☐ |
| Organize saved content into topic folders (e.g., Prompting, Ethics, Tools) | ☐ |
| Set aside 15-20 minutes each week to review saved AI content | ☐ |
| Share or discuss something you've learned with your network | ☐ |
| Misc: | |
| Misc: | |

# Adaptability reflection worksheet

Use this worksheet to assess how adaptable you are in your AI learning journey and plan for continued growth.

## 1. Self-Assessment: Embracing Change:

• When was the last time I explored a new AI tool or method?

_____

_____

• Do I feel excited or overwhelmed when I hear about a new technology?

_____

_____

• What emotions come up when something I just learned is suddenly outdated?

_____

_____

• Do I seek out change, or do I wait until I'm forced to adapt?

_____

_____

## 2. Quarterly Skills Audit:

• Which AI tools am I using regularly? Are they still effective?

_____

_____

• Are there any new tools or platforms I should be exploring?

_____

- What skills have I improved recently?

_____

_____

- What skills are becoming more important in the industry?

_____

_____

- What areas of my knowledge feel outdated or underdeveloped?

_____

_____

## 3. Letting Go of the Old:

List any tools, routines, or beliefs that may be holding you back from growing:

_____

_____

What can you replace them with to become more efficient or innovative?

_____

_____

## 4. My Adaptability Action Plan

This week, I will: _____

This month, I will: _____

Over the next 90 days, I want to improve: _____

_____

_____

# AI Networking Action Plan

Use this worksheet to expand your network, engage with the AI community, and grow faster through collaboration and connection.

## 1. Attend Virtual or Local Events

Search for upcoming AI events, webinars, or meetups you can attend. Note the date and type of event below:

Event Name: _____

Date/Time: _____

Type (Webinar, Meetup, Hackathon, etc.): _____

What I hope to gain or learn from this event:

_____

_____

## 2. Join and Engage in Online AI Communities

Pick at least 2 online communities to join (Reddit, Discord, LinkedIn groups, etc.):

Community 1: _____

Community 2: _____

How I plan to contribute (ask questions, share tips, respond to others):

_____

_____

### 3. Study or Build with Others

Do you have a study buddy or project partner?

Yes / No

Name or Group: _____

Project or topic we are working on:

_____

Meeting schedule (e.g., Weekly Zoom calls): _____

### 4. Track Your Networking Progress

Use the checklist below to keep momentum:

☐ Attended an AI event this month
☐ Joined a new AI community
☐ Asked or answered a question in a forum or group
☐ Shared a helpful resource or insight
☐ Connected with someone new on LinkedIn
☐ Collaborated on a mini project or challenge

# 30-DAY AI CHALLENGE

## Week 1: Foundations & Prompting Basics

| Day | Challenge |
|-----|-----------|
| 1 | Explore ChatGPT or Claude. Ask it to summarize your favorite article. |
| 2 | Write 3 creative prompts and test them. Tweak them to improve results. |
| 3 | Ask an AI to help you brainstorm 10 business, book, or blog ideas. |
| 4 | Use Perplexity or Elicit.org to research a topic you're curious about. |
| 5 | Have ChatGPT role-play a coach or expert to solve a real problem. |
| 6 | Create an AI mind map or outline for a topic you want to learn. |
| 7 | Reflect: What's one AI use case you're excited about? Journal it. |

## Week 2: Content Creation with AI

| Day | Challenge |
|-----|-----------|
| 8 | Use an AI writer (e.g., Jasper, Writesonic) to generate a blog draft. |
| 9 | Use ChatGPT to rewrite or improve an old email or document. |
| 10 | Generate 3 AI images using DALL-E, Midjourney, or Leonardo.ai. |
| 11 | Create a children's story or visual poem using AI text and images. |
| 12 | Use Notion AI to summarize your week or generate a content calendar. |
| 13 | Ask AI to create 3 social media posts on a topic you like. |
| 14 | Combine text + image AI to make a mini digital zine or poster. |

## Week 3: AI in Everyday Life & Productivity

| Day | Challenge |
|-----|-----------|
| 15 | Use AI to create a weekly meal plan or grocery list. |
| 16 | Ask AI to help declutter your schedule or simplify your routine. |
| 17 | Use AI to build a to-do list with priorities and deadlines. |
| 18 | Summarize a podcast or YouTube video using an AI transcription tool. |
| 19 | Create a budget spreadsheet with AI-generated formulas. |
| 20 | Use AI to write a personal letter, journal entry, or affirmation. |
| 21 | Use ChatGPT to simulate a difficult conversation and practice responses. |

## Week 4: Build, Automate, & Reflect

| Day | Challenge |
|-----|-----------|
| 22 | Build a no-code AI app with Glide or Bubble. |
| 23 | Train a basic model with Teachable Machine or Lobe.ai. |
| 24 | Use Zapier or IFTTT to automate one repetitive task. |
| 25 | Explore AI for data insights—analyze a simple dataset with AI. |
| 26 | Create a narrated AI-generated video with Synthesia or Pictory. |
| 27 | Ask AI to quiz you or create flashcards on any subject. |
| 28 | Build a mini portfolio of your best AI creations so far. |
| 29 | Reflect: How has AI helped you? Where do you want to go deeper? |
| 30 | Share one creation or insight on social media or with a friend! |

# CONCLUSION

As we come to the end of this journey together, let's pause for a moment and reflect on how far you've come. When you first picked up this book, perhaps AI felt like an intimidating wall of jargon, technical details, and abstract concepts. My hope has always been to shine a light through that fog—to show you that AI isn't a tool reserved only for coders and scientists, but a powerful ally that you can understand, shape, and use confidently and ethically in your own life. From those early pages where we explored what AI really means, right up to the hands-on strategies and ethical frameworks, you've transformed from a curious beginner into a capable, empowered AI practitioner.

Throughout these chapters, we broke down complex ideas into clear, manageable pieces, always with your real-world needs in mind. We translated technical language into everyday terms and moved step-by-step from foundational knowledge to practical implementation. The result? You now not only grasp what AI is and how it works, but you also have the confidence to apply these tools thoughtfully—without writing a single line of code. That transformation is the heart of this book: demystifying the world of AI so you can lead, create, and succeed in ways that matter to you.

You've learned the essential pillars that shape today's AI landscape: machine learning, neural networks, natural language processing, and the evolution of data-driven technologies. Together, we peeled back the curtain on algorithms and decision-making processes, revealing not just how these systems function, but why they're relevant for modern professionals. By grounding these concepts in real-life examples—from business operations to freelance gig opportunities—your understanding goes beyond theory. It's now anchored in skills you can use every day.

But this wasn't just about the science under the hood. We dove deep into choosing the right AI tools for your work, no matter your industry or role. Whether you're automating repetitive tasks, exploring new ways to generate content, or using AI to streamline collaboration, you now have a toolkit tailored to boost your productivity and expand your earning potential. The strategies and exercises throughout the book equipped you to experiment, iterate, and discover new income streams, all while keeping control and creativity firmly in your hands.

We explored prompt crafting, an often-underestimated skill that lets you communicate clearly with AI systems and get the results you want. This ability alone opens doors to smarter workflows, sharper analysis, and innovative solutions, whether you're drafting emails, pulling insights from data, or customizing AI outputs for clients. Alongside these tactics, we focused on practical advice: evaluating tools for your unique workflow, integrating them without disrupting your daily grind, and measuring success in ways that make sense for your goals.

Of course, no conversation about AI is complete without addressing ethics and responsibility. You've learned not just to embrace AI's power, but also to question its intentions and impacts. We examined privacy, bias, transparency, and the responsibilities that come with adopting new technology. By thinking critically, never just accepting AI's outputs at face value, you become an informed guardian of fairness, accountability, and human-centered progress. Ethics isn't a box to check; it's a lens through which you guide every decision as an AI user.

Along the way, I hope you experienced some meaningful 'aha' moments. Maybe you realized that you don't need to be a coder to harness AI's benefits. Maybe you saw, for the first time, how AI's limitations are as important as its strengths, that it's a partner, not a replacement. Perhaps the biggest revelation was recognizing how much power you have to shape outcomes by simply asking better questions or setting clearer goals. These insights empower you to approach AI not with fear or skepticism, but with balanced optimism, and a readiness to seize new opportunities while guarding against blind spots.

Learning about AI isn't easy, especially when myths and misconceptions are everywhere. So take a moment to truly celebrate what you've accomplished. Not everyone is willing to face something unfamiliar, invest in their own growth, and put in the work to master such a fast-evolving field. Your willingness to dig in, reading, practicing, reflecting, has set you apart. The practical skills you've gained, from navigating interfaces to crafting prompts and evaluating ethical risks, are achievements that will serve you for years to come.

More than anything, you should feel proud of your progress. Remember the nervousness or uncertainty you might have felt at the beginning? Look at you now: ready to bring AI into your workflow, share your expertise with others, and continue building on a solid foundation. If you've completed activities, tried out tools, or even just had a few lightbulb moments along the way, those are milestones worth celebrating. Keep those stories close, they're proof of your dedication, and they might just inspire someone else to start their own journey too.

But don't let your learning stop here. Think of this book as a springboard, a launchpad to the next stage of your development. The world of AI is always moving forward, and staying adaptable is one of your greatest strengths. Consider joining online AI communities or local meetups to stay connected to ongoing trends. Follow the newsletters and podcasts mentioned earlier to keep your finger on the pulse of innovations and challenges. And if you're feeling adventurous, dive into specialized courses or experiment with new applications you find intriguing. Continuous learning keeps you ahead of the curve and prepares you for whatever comes next.

Remember, everything you've achieved becomes even more powerful when you share it. Take what you've learned back to your team, your business, or your network. Start conversations about responsible AI use. Mentor someone who's unsure where to begin or advocate for thoughtful practices in your organization. Share your wins, your lessons, and even your setbacks on social media or in professional forums, because the ripple effect of your actions can drive positive change well beyond your immediate circle.

Building community is key. The more voices contributing to discussions about AI, the more positive, diverse, and ethical our technological future becomes. Your participation helps debunk myths, spread responsible innovation, and ensure that AI works for everyone, not just the few. Whether you're leading workshops, exchanging ideas in a virtual group, or simply supporting colleagues in adopting new tools, know that your involvement matters. Together, we can champion an AI-powered world that values humanity above all.

Thank you for trusting me to be one of your guides in this adventure. I'm deeply grateful for your trust, your time, your curiosity, and your commitment. That says so much about your character and vision for the future. I appreciate every effort you put into reading, questioning, experimenting, and growing. Your experiences enrich this community and help us all learn together. As you move forward, remember that you're part of a movement of leaders, creators, and thinkers shaping the future of AI from the ground up. Our shared mission is bigger than any single book. It's about ensuring technology serves people, not the other way around.

So, what happens now? The next chapter is yours to write. Approach AI, and the changes it brings, not as obstacles, but as opportunities for growth, creativity, and impact. You now have both the mindset and the practical skills to navigate this exciting landscape with confidence and care. Lead by example, challenge assumptions, and never stop asking how AI can help you and those around you live and work better. The future you're helping to build is bright, promising, and powered by your informed, ethical choices. Go forward with courage, curiosity, and the conviction that you can and will make a difference.

# REFERENCES

Russell, S. J., & Norvig, P. (2020). Artificial Intelligence: A Modern Approach (4th ed.). Pearson.

Turing, A. M. (1950) Computing Machinery and Intelligence. Mind 49: 433-460.

University Of San Diego, Strategic Plan, Envisioning 2024

Sarker, I.H. (2021) Machine Learning: Algorithms, Real-World Applications and Research. Directions. SN Computer

Science, 2, Article #160 .https://doi.org/10.1007/s42979-021-00592-x

Saleem, S. (2023, May 15). *Neural Networks in 10mins. Simply Explained!* Medium. https://medium.com/@sadafsaleem5815/neural-networks-in-10mins-simply-explained-9ec2ad9ea815

McCullum, N. (2020, June 28). *Deep Learning Neural Networks Explained in Plain English.* FreeCodeCamp.org. https://www.freecodecamp.org/news/deep-learning-neural-networks-explained-in-plain-english/

Frank. (2025, May). *The Honest Truth About What AI Can (and Can't) Do.* Trust Consulting Services. https://www.trustconsultingservices.com/ai-honest-truth-can-and-cant-do/

*AI Myths Debunked: True & Interesting Facts about AI – 365 Data Science.* Blog: https://365datascience.com/trending/ai-myths-debunked/

*The CIO's Guide to Avoid AI-Washing: 9 Tips for Vetting AI Vendors and Solutions - On-Demand Group.* (2024, June 18). On- Demand Group. https://www.ondemandgroup.com/the-cios-guide-to-avoid-ai-washing-9-tips-for-vetting-ai-vendors-and-solutions/

Radanliev, P. (2025, February 7). *AI Ethics: Integrating Transparency, Fairness, and Privacy in AI Development.* Applied Artificial Intelligence; Informa UK Limited. https://doi.org/10.1080/08839514.2025.2463722

Singhal, A., Neveditsin, N., Tanveer, H., & Mago, V. (2024, April 3). *Toward Fairness, Accountability, Transparency, and Ethics in AI for Social Media and Health Care: Scoping Review.* JMIR Medical Informatics. https://doi.org/10.2196/50048

Capitol Technology University. (2023, May 30). *The Ethical Considerations of Artificial Intelligence.* https://www.captechu.edu/blog/ethical-considerations-of-artificial-intelligence

Pazzanese, C. (2020, October 26). *Great promise but potential for peril.* Harvard Gazette; Harvard University. https://news.harvard.edu/gazette/story/2020/10/ethical-concerns-mount-as-ai-takes-bigger-decision-making-role/

Kimbrough, K. (2025, January 21). *AI Is Shifting the Workplace skillset. but Human Skills Still Count.* World Economic Forum. https://www.weforum.org/stories/2025/01/ai-workplace-skills/

Kannan SP. (2024, October 15). *Emotional Intelligence and AI: Adapting to New Challenges.* Adyog | Creative Design and Digital Product Development Company. https://blog.adyog.com/2024/10/15/emotional-intelligence-and-ai-adapting-to-new-challenges/

Rebelo, M. (2023, September 1). *The 5 best no-code app builders in 2023 | Zapier.* Zapier.com. https://zapier.com/blog/best-no-code-app-builder/

*User Onboarding: Top 32 Tools [Categorized & Updated].* (2025). Userguiding.com. https://userguiding.com/blog/user-onboarding-tools

Grigoryan, S. (2025, March 17). *AI User Onboarding: 7 Strategies to Automate Customer Onboarding.* Thoughts about Product Adoption, User Onboarding and Good UX | Userpilot Blog; Userpilot. https://userpilot.com/blog/ai-user-onboarding/

Rebelo, M. (2023, August 1). *The 6 best AI scheduling assistants in 2023 | Zapier.* Zapier.com. https://zapier.com/blog/best-ai- scheduling/

Team, S. (2025, May 21). *40 AI task automation tools | The definitive list.* Superhuman Blog. https://blog.superhuman.com/ai-task-automation-tools/

Ivanauskaite, M. (2025, June 11). *20 Easy Side Hustle Ideas from Home for Beginners | JumpTask.* Jumptask.io; JumpTask. https://jumptask.io/blog/side-hustle-ideas-from-home/

Reffell, C. (2021, February 26). *How To Earn an Extra Income Through Top Crowdsourced Microtasking Platforms.* Crowdsourcing Week. https://crowdsourcingweek.com/blog/how-to-earn-an-extra-income-through-top-crowdsourced-microtasking-platforms/

Akbar, S. (2025, April 7). *Techugo.* Techugo.com. https://www.techugo.com/blog/ai-personal-assistant-apps-transforming-everyday-productivity/

Kadambari Darad. (2024, December 20). *Understanding Adaptive Learning: How AI Is Revolutionizing Personalized Education.* ELearning Industry. https://elearningindustry.com/understanding-adaptive-learning-how-ai-is-revolutionizing-personalized-education

Karl, T. (2024, March 30). *Revolutionizing Workplaces: Real-World Examples of AI Implementation.* New Horizons. https://www.newhorizons.com/resources/blog/examples-of-ai-in-the-workplace

Arhip, B. (2024, October 10). *10 Use Cases of AI in HR with real-world case studies.* Cubeo. https://www.cubeo.ai/10-use-cases-of-ai-in-hr-with-real-world-case-studies/

Baveling. (2024, October 26). *AI in Everyday Life: Discovering the Unseen Influence on Our Daily Routines.* https://www.baveling.com/post/ai-in-everyday-life-discovering-the-unseen-influence-on-our-daily-routines

Jamil Valliani. (2024, September 18). *The ultimate guide to writing effective AI prompts - Work Life by Atlassian.* Work Life by Atlassian. https://www.atlassian.com/blog/artificial-intelligence/ultimate-guide-writing-ai-prompts

MIT Sloan Teaching & Learning Technologies. *Effective Prompts for AI: The Essentials.* (n.d.). https://mitsloanedtech.mit.edu/ai/basics/effective-prompts/

The AI Enterprise. (2025). *Prompt Engineering Deep Dive.* The Artificially Intelligent Enterprise. https://www.theaienterprise.io/p/2025-prompt-engineering-update

Caltech.edu. *AI Tools for Everyone: A Hands-On Learning Lab (Online).* (2025). https://ctme.caltech.edu/artificial-intelligence-tools-for-everyone-online.html

Isael Paz-Zimbeck. (2025, March 21). *Build Gen AI Confidence for Workforce Agility & Productivity* - Degreed. https://explore.degreed.com/blog/build-gen-ai-confidence-for-workforce-agility-productivity/

Siu, E. (2024, March 20). *Single Grain, LLC.* Single Grain. https://www.singlegrain.com/blog/a/ai-skills-resources/

DeepLearning.AI. (n.d.). The Batch – Weekly AI news. https://www.deeplearning.ai/the-batch/

Clark, J. (n.d.). Import AI – Newsletter. https://jack-clark.net/

AI Weekly. (n.d.). AI Weekly Newsletter. https://aiweekly.co/

Cognilytica. (n.d.). AI Today Podcast. https://www.cognilytica.com/aitoday/

Changelog Media. (n.d.). Practical AI Podcast. https://changelog.com/practicalai

Fridman, L. (n.d.). The Lex Fridman Podcast. https://lexfridman.com/podcast/

Reddit. (n.d.). r/MachineLearning Subreddit. https://www.reddit.com/r/MachineLearning/

Reddit. (n.d.). r/learnmachinelearning Subreddit. https://www.reddit.com/r/learnmachinelearning/

Google. (n.d.). Teachable Machine. https://teachablemachine.withgoogle.com/

Canva. (n.d.). Magic Studio. https://www.canva.com/magic/

Runway. (n.d.). AI video editing tools. https://runwayml.com/

Pictory. (n.d.). AI video generator. https://pictory.ai/

Meetup. (n.d.). Find events. https://www.meetup.com/

Eventbrite. (n.d.). Discover AI events. https://www.eventbrite.com/

Pomofocus.io. (n.d.). Pomodoro timer for productivity. https://pomofocus.io/

Habitica. (n.d.). Gamified habit tracker. https://habitica.com/

Notion. (n.d.). All-in-one workspace.

Trello. (n.d.). Visual task management tool. https://trello.com/

www.ingramcontent.com/pod-product-compliance
Lightning Source LLC
La Vergne TN
LVHW021348080426
835508LV00020B/2162